Atlas of sedimentary rocks under the microscope

Atlas of sedimentary rocks under the microscope

A.E. Adams, W.S. MacKenzie and C. Guilford

Addison Wesley Longman Limited
Edinburgh Gate
Harlow
Essex CM20 2JE, England
and Associated Companies throughout the world

First published 1984
Second impression 1987
Third impression 1991
Fourth impression 1994
Fifth impression 1995
Sixth impression 1997

British Library Cataloguing in Publication Data

Adams, A. E.
Atlas of sedimentary rocks under the microscope.
1. Rocks, sedimentary – Pictorial works
I. Title II. MacKenzie, W. S.
III. Guilford, C.
552'.5'0222 QE471

ISBN 0-582-30118-1

Library of Congress Cataloging in Publication Data
A catalog entry for this title is available from the Library of Congress.

Produced by Longman Asia Limited, Hong Kong
GC/06

Contents

Preface

The study of rocks using thin sections and a petrographic microscope was initiated by Henry Clifton Sorby in the middle of the nineteenth century and the first rocks he described were silicified limestones from the Jurassic in Yorkshire. This work was published in 1851. His presidential address to the Geological Society of London in 1879 was entitled 'On the structure and origin of limestones' and Sorby had a series of plates, made from camera lucida drawings, reproduced for private circulation with copies of the text of his address. These illustrated the microscopic characteristics of limestones from throughout the British geological record and amounted to the first petrographic atlas.

Despite the pertinence of Sorby's work, much of which is still valid today, few people recognized its importance at the time. While the petrographic study of igneous and metamorphic rocks became increasingly important, that of sedimentary rocks languished until well into the present century. Since about 1950, with much geological research directed towards the search for oil and gas trapped in the pore-spaces of sedimentary rocks, sedimentary petrography has become one of the most important fields of geology and forms a key part of most undergraduate courses.

The aim of this book is therefore similar to that of the previously published *Atlas of igneous rocks and their textures*, in that it is designed to be a laboratory handbook for the student beginning a study of sedimentary rocks in thin section, whether he or she is an amateur or an undergraduate. Only a basic knowledge of mineralogy and palaeontology is assumed. While we make no claim that the book is comprehensive, we have tried to include photographs of most of the components of sedimentary rocks encountered in thin sections during an undergraduate course in geology.

The book is in three parts. Part 1 deals with the terrigenous clastic rocks and concentrates on sandstones, since the petrographic microscope is most usefully employed with rocks of this grain size. We have attempted to show the common detrital components of sandstones and the range of rock types occurring, without becoming involved in details of the many classifications which exist.

Part 2 deals with the carbonate rocks and is the longest section in the book. This is because to the newcomer to carbonate petrology, limestones contain a bewildering variety of grain types. The bioclasts in particular show such variation in shape and structure that it has been difficult to know what to leave out. We have attempted to show the range of common bioclast types while realizing that this section of the book cannot be comprehensive within the limits of the number of photographs which we are able to reproduce. Most of the photographs of limestones are from stained thin sections and acetate peels. The staining aids identification of minerals and textures and also makes limestones more attractive to study. The reader examining a collection of unstained sections of carbonate rocks should still find the photographs and text useful in identifying grain types and textures. Photographs of unstained limestone sections are included throughout to remind the reader what untreated material looks like.

Part 3 illustrates ironstones, cherts, evaporites, phosphorites and carbonaceous rocks in thin section. We hope the section on evaporites will be of particular interest, as published colour photomicrographs of some minerals are rare.

Three appendices are included. Appendix 1 is a slightly modified form of the appendix in the *Atlas of igneous rocks and their textures* and describes how a thin section may be made. Appendix 2 describes a method of staining thin sections of limestones and Appendix 3 contains instructions on how to make acetate peels.

Throughout the book we have tried to keep the text descriptive and to avoid details of interpretation. However, it has proved impossible to omit discussion in some cases, particularly with the carbonate rocks where identification of grains and textures goes hand in hand with an interpretation of their origin. We have attempted to show typical material rather than particularly good examples of any feature illustrated. Extensive cross-referencing is given to help the reader in finding other photographs of similar phenomena.

Inevitably the bulk of the illustrated material comes from the British Isles; we believe however that it is representative of sedimentary rocks the world over.

Finally, we must repeat the cautionary note in the preface to *Atlas of igneous rocks and their textures*. This book is a laboratory handbook to *assist* in the study of sedimentary rocks in thin section. There is no substitute for the student examining material under the microscope for him- or herself and we hope this book will encourage students to make their own petrographic observations.

Acknowledgements

Although this book is based on thin sections and acetate peels held in the teaching collections of the Department of Geology, University of Manchester, it would not have been possible without the generous loan of material from the research collections of many colleagues. We are particularly indebted to Professor Sir Frederick Stewart who loaned much of the material for the evaporites section. We are grateful to Drs. J. M. Anketell, P. Gutteridge, J. Kantorowicz, J. E. Pollard. A. T. S. Ramsay, K. Schofield, Mr R. D. Vaughan and Professor E. K. Walton, all of whom loaned material and made suggestions or comments on the manuscript. We would also like to thank Professor J. B. Dawson for permission to include a photograph of one of Sorby's thin sections from the collection held at Sheffield University.

We wish to thank Patricia Crook for her patient typing of various versions of the text and Phil Stubley for drafting the originals of the diagrams. Finally we wish to acknowledge the help given to us by all the staff of the Longman Group.

We acknowledge permission from Springer Verlag and Professor F. J. Pettijohn to reproduce Figs. A and D, and the American Association of Petroleum Geologists for Figs. E and F and Tables 3 and 4.

Part 1

Terrigenous clastic rocks

Introduction

Terrigenous clastic sediments are made up of transported fragments derived from the weathering of pre-existing igneous, sedimentary or metamorphic rocks. These rocks are classified initially according to grain size, using the Udden-Wentworth scale (Table 1).

It is those terrigenous sediments of intermediate grain size – the coarser siltstones, sandstones and finer conglomerates and breccias – that are most usefully studied using the petrographic microscope, since the grain types can be identified by this means. The principal component grain types are quartz, feldspar and rock fragments. The matrix of such sediments may be the fine-grained weathering products of the source rocks, such as clay minerals, or it may be a secondary cement.

Clays and shales are too fine-grained for study using the petrographic microscope and must be examined by electron microscopy or X-ray diffraction. The components of coarser conglomerates and breccias can usually be identified with the aid of only a hand lens.

The shape and roundness of the components of terrigenous clastic rocks are important in describing sedimentary textures. Categories of roundness for grains of high and low sphericity are shown in Fig. A. Sedimentary textures are discussed on p. 24.

Table 1. Grain-size classification of sediments

Size in mm of class boundary	Class term	Grain size terms for rock	
	boulders	rudite rudaceous rock conglomerate breccia	
256	cobbles		
64	pebbles		
4	granules		
2	very coarse sand	arenite arenaceous rock sandstone	
1	coarse sand		
0.5($\frac{1}{2}$)	medium sand		
0.25($\frac{1}{4}$)	fine sand		
0.125($\frac{1}{8}$)	very fine sand		
0.0625($\frac{1}{16}$)	coarse silt	siltstone	argillite argillaceous rock mudstone mudrock shale
0.0312($\frac{1}{32}$)	medium silt		
0.0156($\frac{1}{64}$)	fine silt		
0.0078($\frac{1}{128}$)	very fine silt		
0.0039($\frac{1}{256}$)	clay	claystone	

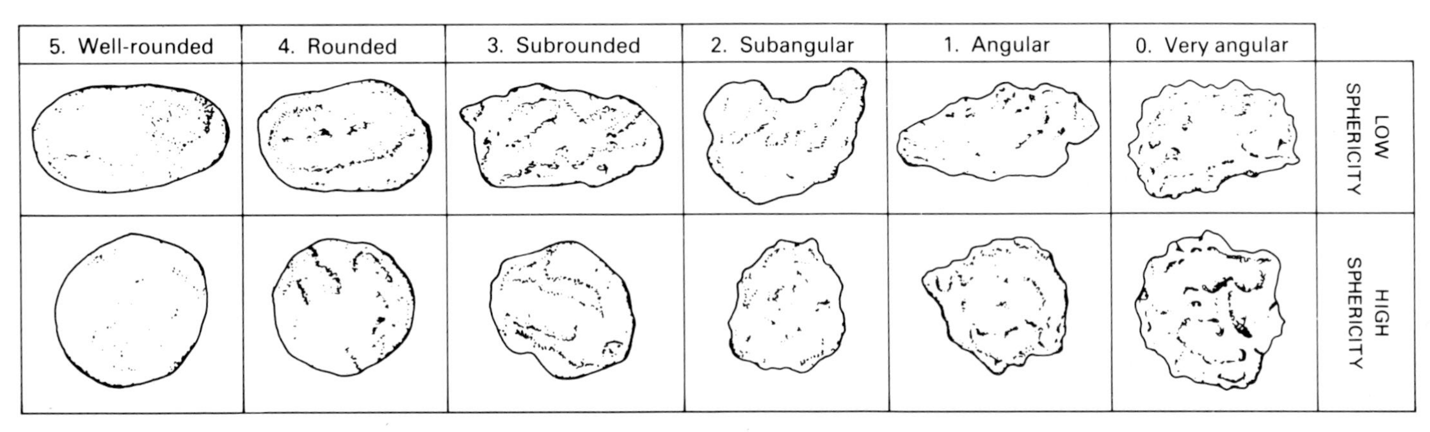

Fig. A Categories of roundness for grains of low and high sphericity (after Pettijohn et al., 1973)

Quartz

The most abundant grain type in sandstones and conglomerates is quartz. In addition to the size and shape of individual quartz grains, the following features should be observed since they may provide clues to the provenance of the sediment:

1. Whether the quartz grains are single crystals (monocrystalline) or are made up of a number of crystals in different orientations (polycrystalline).
2. Whether extinction is uniform (the grain extinguishes in one position on rotation of the stage) or undulose (the grain extinguishes over a range of at least 5° on rotation of the stage).
3. The presence or absence of inclusions.
4. In the case of polycrystalline grains, whether the crystal boundaries are straight or sutured.

1 and **2** show subrounded quartz grains which are single crystals, taken with plane-polarized light (PPL) and with crossed polars (XPL). The matrix between the sand grains contains opaque iron oxide and some calcite. The latter shows high-order pink and green interference colours.

*1 and 2: Red Mountain Formation, Silurian, Birmingham, Alabama, USA; magnification × 38; **1** PPL, **2** XPL.*

Quartz

(continued)

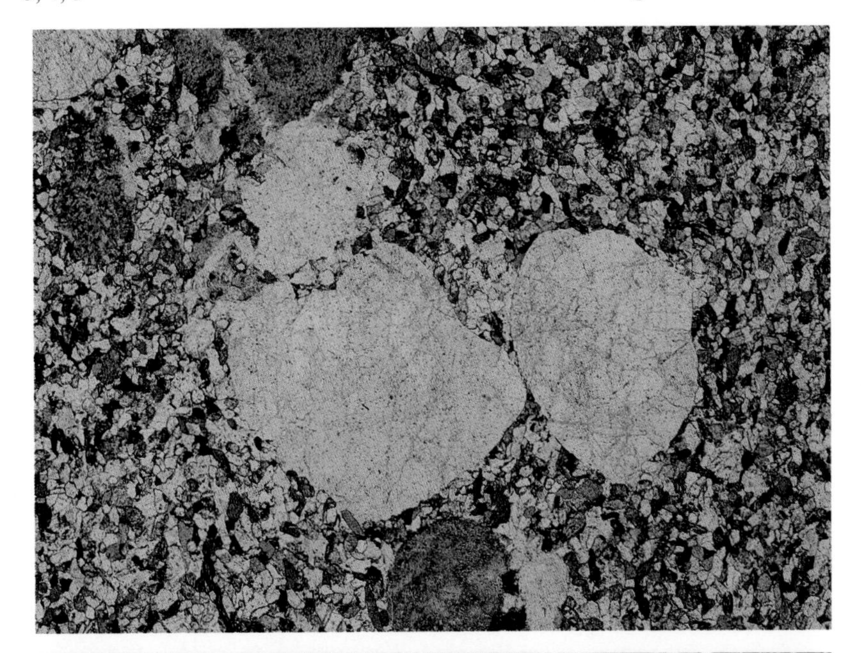

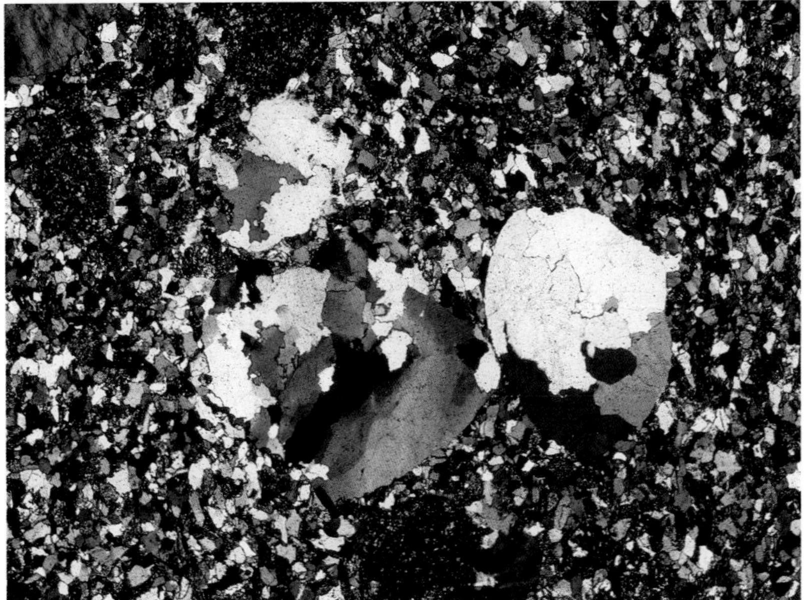

The three rounded grains in the centre of **3** and **4** are made up of a number of quartz crystals in different orientations and are thus *composite* or *polycrystalline* quartz. The composite nature of the grains is clear only in the view taken with polars crossed. Note that the boundaries between the crystals are sutured. This is characteristic of quartz from a metamorphic source. Composite quartz from igneous sources usually has straighter crystal boundaries. The much finer sediment surrounding the composite quartz grains contains monocrystalline quartz and brownish clasts of fine-grained material which are probably shale or slate fragments.

5 shows a composite quartz grain viewed under crossed polars, in which not only are the crystal boundaries within the grain sutured, but also the crystals are elongated in a preferred direction. Such grains are called *sheared* quartz or *stretched metamorphic* quartz. In this type of quartz, individual crystals normally show undulose extinction as a result of strain. Evidence for this in the example shown comes from the non-uniform interference colours shown by many of the crystals.

*3 and 4: Trichrug Beds, Silurian, Pontarllechau, Dyfed, Wales; magnification × 16; **3** PPL, **4** XPL.*
5: Carboniferous, Anglesey, Wales; magnification × 43, XPL.

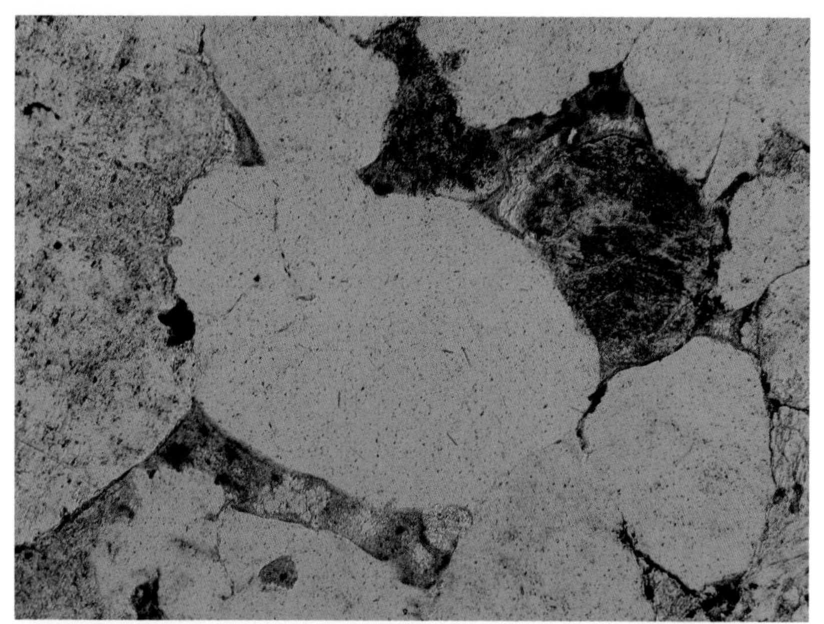

Quartz

(continued)

The quartz grain in the centre of the field of view in **6** appears to be a single homogeneous crystal. In **7** however, where the same field of view is seen under crossed polars, the quartz grain is clearly made up of parts of two crystals. One, comprising the upper left portion of the grain is showing a mid-grey interference colour, whereas the rest of the grain comprises a crystal with areas showing slightly different interference colours. The left- and right-hand sides are in extinction and interference colours become progressively paler towards the centre of the grain. Such a grain would show sweeping extinction when rotated. This phenomenon, known as *undulose extinction*, is a result of strain and is found in quartz grains from both igneous and metamorphic sources.

Quartz crystals may sometimes incorporate mineral inclusions and identification of the minerals may yield information about the provenance of the sediment. The quartz grain in the centre of **6** and **7** has a number of needle-shaped inclusions, although they are too small for the mineral to be identified at the magnification shown. Inclusions of the fluid present at the time of crystallization are common in quartz crystals and are known as *fluid inclusions* or *vacuoles*. **8** shows a quartz grain with abundant vacuoles. These appear as dark specks, and in the sample illustrated, many are concentrated in lines running at a low angle to the length of the picture. Quartz with abundant vacuoles is usually derived from a source of low-temperature origin, such as a hydrothermal vein, and appears milky-white in a hand specimen. the photograph also shows a green mineral in the matrix around the quartz grain, which is chlorite.

6 and 7: locality and age unknown; magnification × 72; ***6*** *PPL, 7 XPL.*
8*: Coal Measures, Upper Carboniferous, Lancashire, England; magnification × 72, PPL.*
Undulose extinction can also be seen in **5**.

Feldspar

Feldspars are a major constituent of many sandstones and conglomerates. Alkali feldspars are more common than calcic plagioclase, partly because they are more resistant to chemical weathering, and partly because the ultimate source of many terrigenous rocks is granite or gneiss, rocks in which the feldspars are mainly the alkali varieties. The chemical weathering of feldspars may be rapid, producing micas and clay minerals. Therefore feldspars are most abundant and best preserved in rocks derived from mechanical weathering. The identification of feldspars in thin section is straightforward in the case of multiple-twinned grains of plagioclase or microcline, or where perthitic textures are present. Distinguishing between untwinned orthoclase and quartz can be difficult. The following features may help:

1. Alteration – because orthoclase is more susceptible to chemical weathering than quartz, it is often cloudy or brown-coloured in PPL, whereas quartz is usually clear.
2. Refractive Index – the index of quartz is very close to, but higher than that of Canada balsam, whereas the index of orthoclase is always lower than balsam.
3. Interference figure – orthoclase is biaxial with a moderate 2V, quartz is uniaxial unless strained.

9 and **10** show a large plagioclase grain which is easily identified by the twinning in the photograph with polars crossed. The grain shows a combination of two types of twins which are probably Carlsbad (simple twin) and albite (multiple twinning). The cloudiness seen in PPL is caused by patchy alteration of the feldspar. The highly birefringent, fine-grained alteration product is probably sericite, a mica.

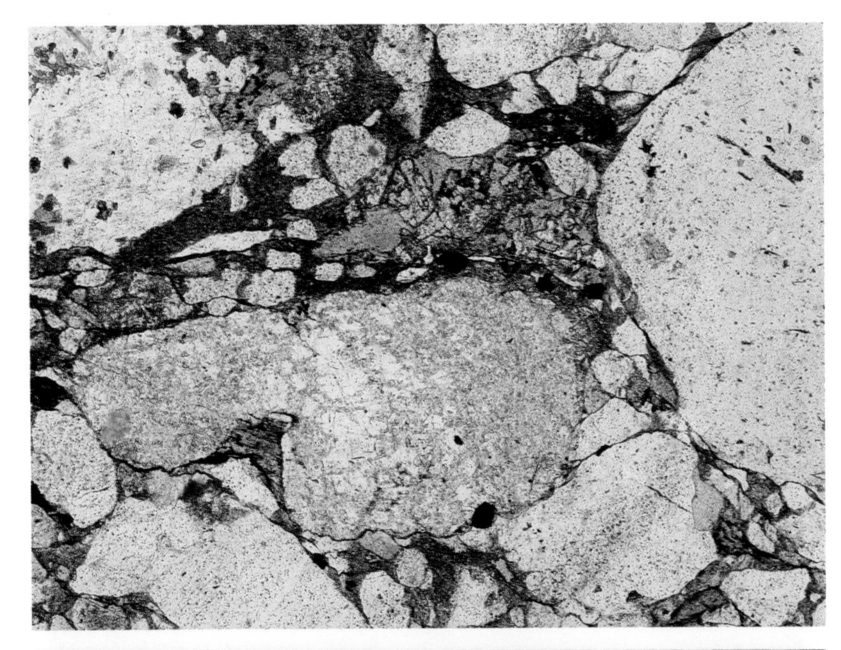

***9** and **10**: Caban Conglomerate, Silurian, Rhayader, Wales; magnification × 25; **9** PPL, **10** XPL.*

Feldspar

(continued)

11 and **12** show a pebble-sized fragment composed almost entirely of microcline. Microcline can be identified easily by the cross-hatched twinning which it invariably shows. Although the microcline shows little alteration, feldspar grains in the upper left, including multiple-twinned plagioclase, are brownish coloured as a result of alteration. In contrast, the quartz in the upper right is relatively clear and unaltered.

Grains showing perthitic intergrowths, comprising blebs or lamellae of sodium-rich feldspar in potassium-rich feldspar, are not uncommon in sediments. **13** and **14** show a very coarse sand-sized fragment of perthite. Most of the other sediment grains are quartz and the matrix contains highly birefringent mineral grains too small to identify at the magnification shown.

15 and **16** show grains of orthoclase and quartz. The feldspar can be identified in the PPL view by its cloudy appearance due to alteration. The quartz is clear and unaltered. In the photograph taken with polars crossed, it can be seen that one feldspar exhibits a simple Carlsbad twin (upper right of photograph), but most of the grains are not twinned. Two multiple-twinned plagioclase crystals are also visible in the field of view.

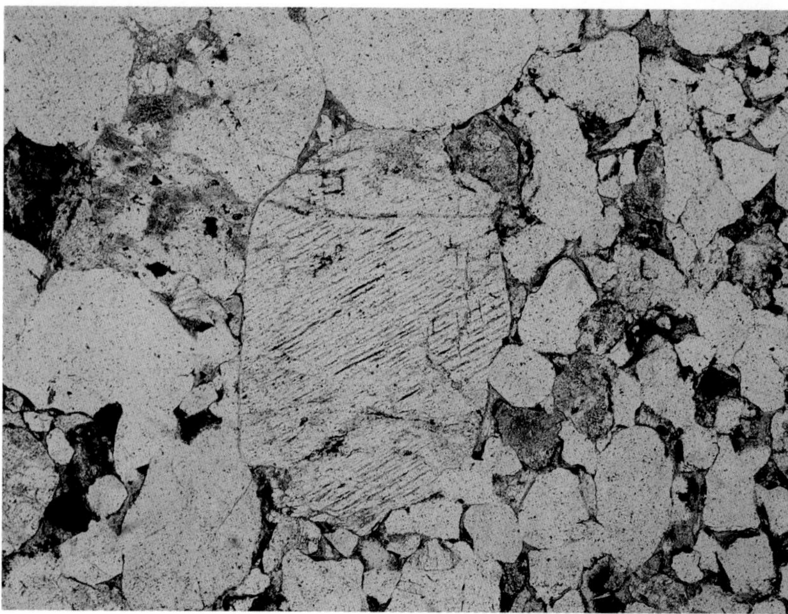

Feldspar

(continued)

*11 and **12**: Torridonian, Precambrian, Isle of Skye, Scotland; magnification × 16; **11** PPL, **12** XPL.*
*13 and **14**: Torridonian, Precambrian, Torridon, Scotland; magnification × 36; **13** PPL, **14** XPL.*
***15** and **16**: Torridonian, Precambrian, Scotland; magnification × 72; **15** PPL, **16** XPL.*
*Feldspars are also shown in **54**, **55**, **58**, **59**, **64** and **65**.*

Rock fragments

Rock fragments, in particular metamorphic rock fragments, are important contributors to many detrital sediments.

17 and **18** show a sediment with many rock fragments. The two fragments in the centre of the photograph above the large quartz grain are made up of fine-grained material which cannot be resolved at this magnification. They are fragments of shale or slate, and the characteristic platy shape is a result of derivation from a cleaved source rock containing abundant platy minerals. The sediment is very poorly-sorted, containing many small rock frgments, quartz grains and at least one twinned feldspar (in the centre, near the top), as well as the large quartz grain, part of which is seen at the base of the photograph.

Fragments of coarser-grained metamorphic rocks are often schistose. **19** and **20** show a fragment of muscovite-bearing quartz-rich rock. The mica flakes show a preferential alignment resulting in a schistose texture. Such fragments are sometimes classified as *schistose quartz* rather than metamorphic rock fragments.

Sedimentary rock fragments, other than chert, are relatively uncommon in terrigenous sedimentary rocks because they usually break down fairly easily into their component grains. **21** and **22** show a large sandstone fragment. Note that although the component particles are all quartz, they are clearly distinguishable even with PPL. This contrasts with the composite quartz grain shown in **3** and **4**, where individual crystals are not visible in PPL. The photograph taken with XPL shows that the individual quartz grains are separated by a cement with bright interference colours. This is likely to be clay.

Rock fragments

(continued)

***17** and **18**: Caban Conglomerate, Silurian, Rhayader, Wales; magnification ×27; **17** PPL, **18** XPL.*
***19** and **20**: Caban Conglomerate, Silurian, Rhayader, Wales; magnification ×28; **19** PPL, **20** XPL.*
***21** and **22**: Arenig Conglomerate, Rhosneigr, Anglesey, Wales; magnification ×16; **21** PPL, **22** XPL.*

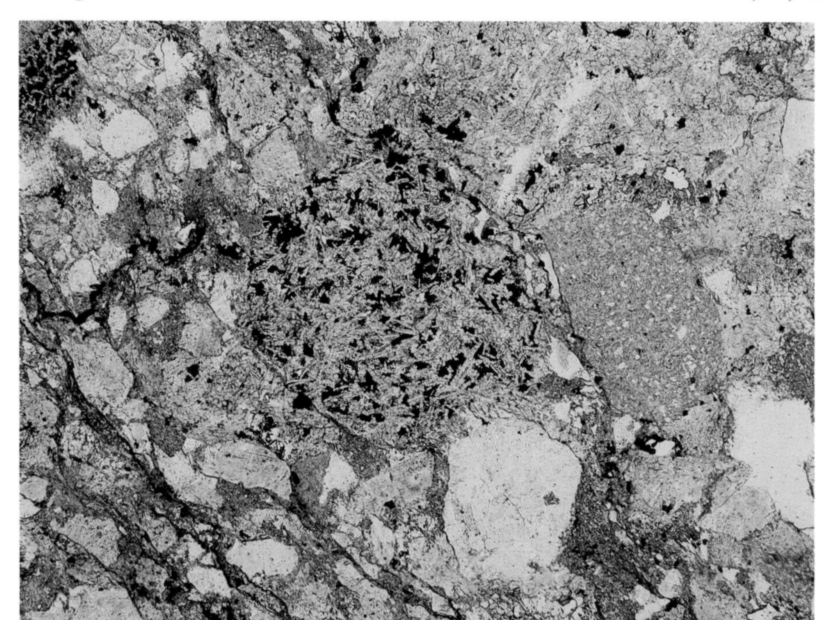

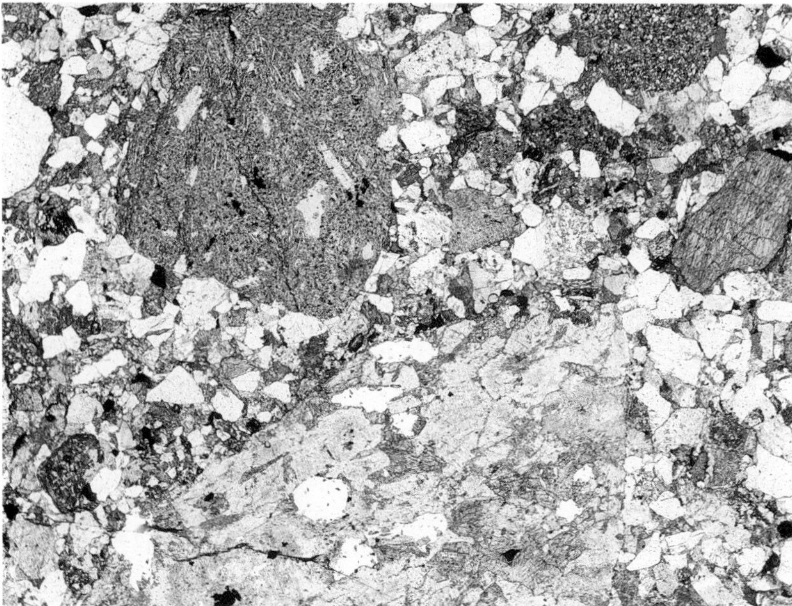

Rock fragments

(continued)

The variety of igneous rock fragments found in sediments is as great as the variety of igneous rocks themselves and lack of space prevents the inclusion of more than a few examples. Reference to the *Atlas of igneous rocks and their textures* may assist in the identification of fragments. The ferro-magnesian minerals which are common in basic igneous rocks are relatively unstable in earth surface conditions and often show alteration, making identification of fragments difficult.

23 and **24** show a volcanic rock fragment in the centre of the field of view. It consists of plagioclase laths set in an altered groundmass which is too fine-grained for its constituents to be identified at the magnification shown. A second rock fragment to the right of centre is composed of quartz crystals set in a birefringent matrix probably of clay minerals. The sediment also contains separate feldspar grains, some of which are multiple-twinned plagioclase, and both monocrystalline and polycrystalline quartz. The matrix of the whole rock contains birefringent clay or mica minerals.

25 and **26** show two different igneous rock fragments. To the left and above the centre of the field of view is a fine-grained, probably volcanic, basic rock. It consists of microphenocrysts of plagioclase feldspar set in a groundmass of feldspar, very small pyroxene crystals and opaques. Pale green chlorite occurs, possibly filling original vesicles. This chlorite is black in the XPL view owing to its very low birefringence. The lower part of the field of view is mostly occupied by a coarse-grained plutonic rock fragment consisting mainly of plagioclase feldspar and pyroxene. The simple-twinned feldspars may be alkali feldspar, although in this case no difference in refractive index between them and the multiple-twinned plagioclase feldspars could be detected. This illustrates the difficulty in precise identification of igneous rock fragments. In addition to smaller rock fragments the sediment contains subangular quartz grains and on the right hand edge a single crystal of a ferro-magnesian mineral, probably amphibole, showing an orange interference colour in the XPL view. The relatively fresh igneous rock fragments and ferro-magnesian mineral grains suggest that this sediment underwent little transport after erosion from source rocks.

Chert fragments are quite common in sedimentary rocks since chert is stable and resistant to weathering. Plates **27** and **28** show a thin section of a conglomerate in which the large rounded fragments are chert. The view

Rock fragments

(continued)

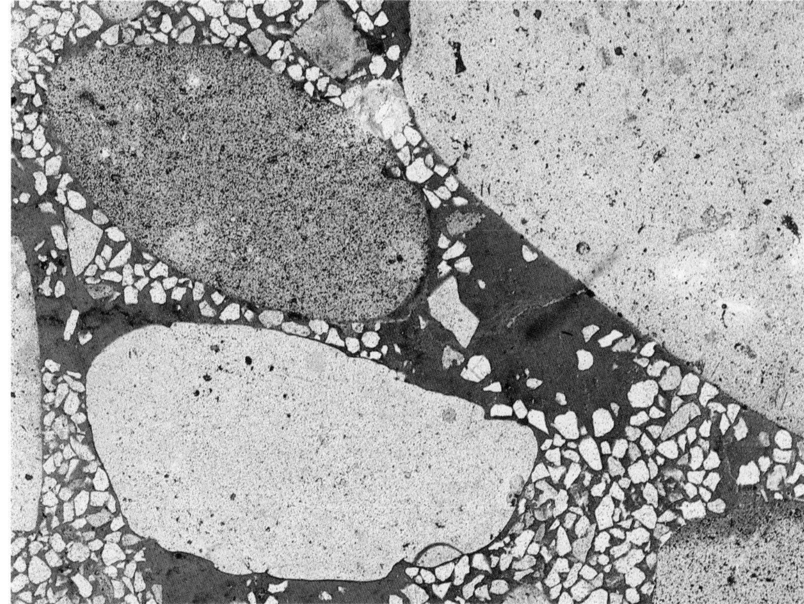

taken with crossed polars shows that the fragments are made up of very fine-grained quartz (micro-quartz, p. 82). Small fragments of chert can be difficult to distinguish from fine-grained acid volcanic rocks although the latter may show porphyritic textures. In the sample illustrated the matrix contains subangular to subrounded quartz grains and small chert fragments set in an iron oxide-rich cement (brown in PPL).

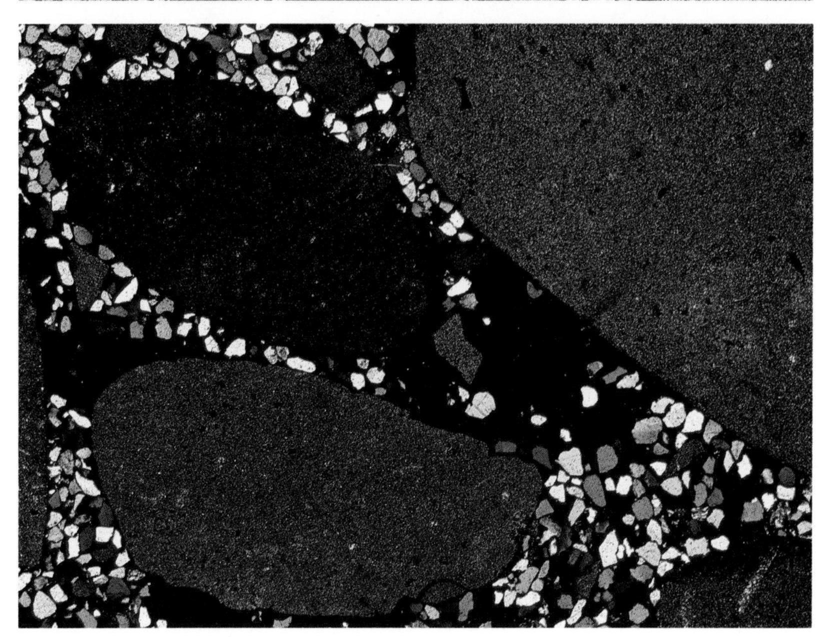

***23** and **24**: Gwstadnant Grit, Ordovician, Clogwyn, Gwynedd, Wales; magnification × 25; **23** PPL, **24** XPL.*
***25** and **26**: Glen App Conglomerate, Ordovician, Ayrshire, Scotland; magnification × 11; **25** PPL, **26** XPL.*
***27** and **28**: Hertfordshire Puddingstone, Tertiary, Chiltern Hills, England; magnification × 13; **27** PPL, **28** XPL.*
Other rock fragments are shown in **3**, **4**, **33**, **34**, **56–61**, **66**, **67**.

Micas

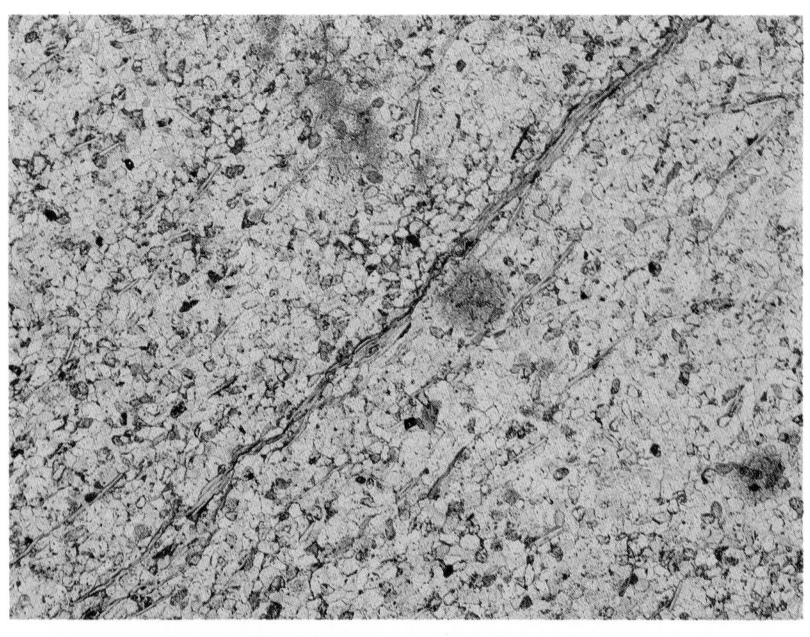

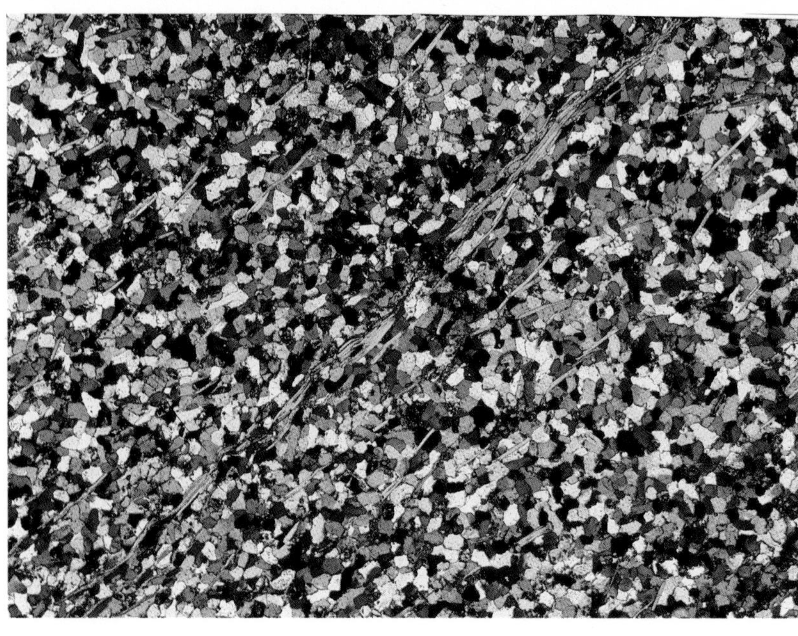

Micas rarely form more than a few percent of any terrigenous sediment, even though they may be conspicuous in hand specimens. Muscovite is more common than biotite since it is more resistant to weathering. The source is usually a granitic or schistose rock. **29** and **30** show a typical micaceous sandstone. Note the parallel alignment of the muscovite flakes and the concentration at a particular level. This indicates the bedding. The thin section has been positioned to show the second-order interference colours of the muscovite. If the thin section was shown with the bedding horizontal, the muscovite flakes would be at extinction.

29 and 30: Tilestones, Silurian, Llangadog, Dyfed, Wales: magnification × 16; 29 PPL, 30 XPL.
Other micas are shown in **68** *and* **69**.

Clay minerals

Clay minerals form a significant fraction of sandstones and are the major constituents of argillaceous rocks. They may be detrital or authigenic. However, since they cannot be readily identified using an ordinary light microscope, but are studied by use of the electron microscope and by X-ray diffraction, they are not considered in detail here.

Clay minerals can be seen in plates **22–24**, **45**, **46**, **62–67**.

Chlorite

The sheet silicate mineral, chlorite, is abundant in sedimentary rocks. It may occur as detrital flakes, usually derived from low-grade metamorphic rocks, as an alteration product, especially of volcanic rock fragments or as an authigenic mineral filling pore-spaces. Plates **31** and **32** show a fine-grained sedimentary rock in which many small fragments are visible, but are less than 1 mm across at this magnification. The rock is therefore a siltstone. The larger rounded grains which are colourless in PPL, and show slightly anomalous bluish-grey interference colours in XPL, are chlorite. In this case chlorite has grown in the rock as a result of the breakdown of small rock fragments and fine-grained matrix during low-grade metamorphism.

*31 and 32 Ordovician, Llangranog, Dyfed, Wales: magnification × 72, **31** PPL, **32** XPL.*

*Chlorite is shown also in **8**, **58** and **59***

Glauconite

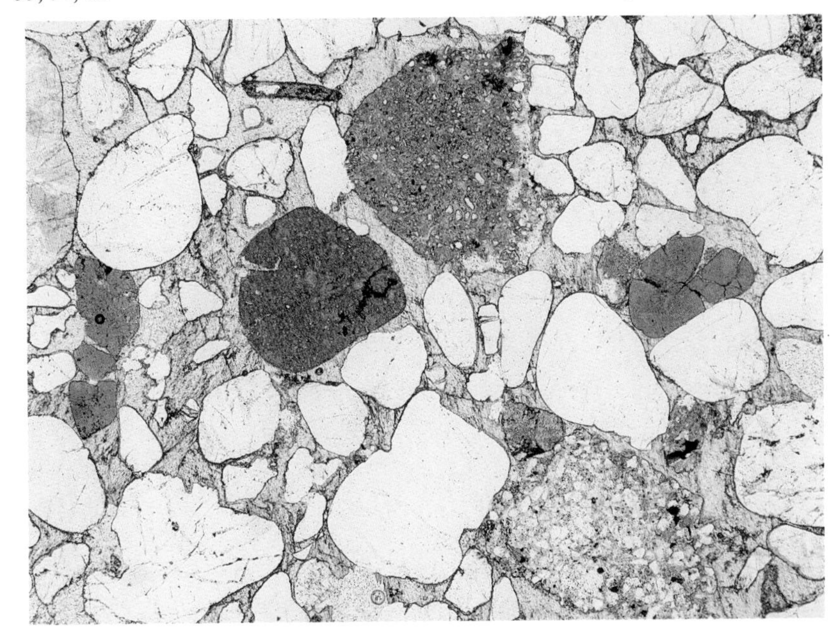

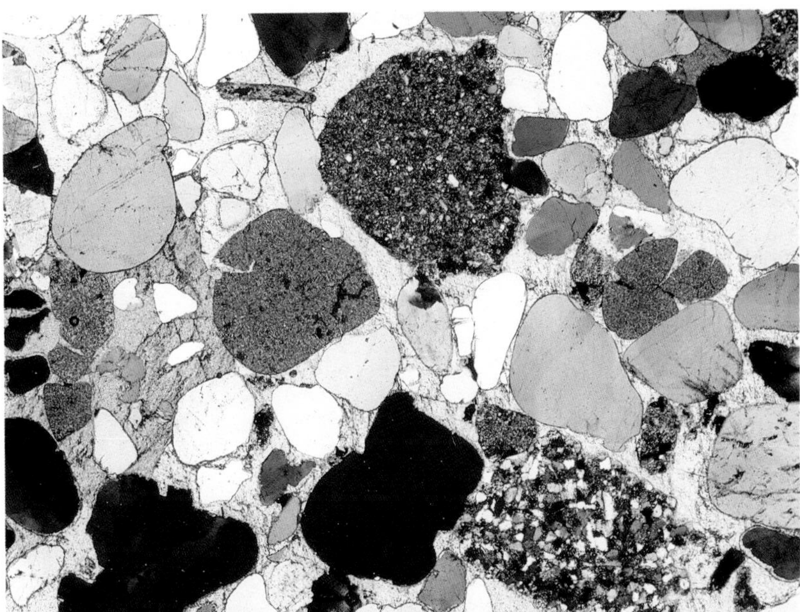

Glauconite is a hydrous potassium iron alumino-silicate mineral which forms exclusively in marine environments, usually in fairly shallow waters. It commonly occurs as rounded pellets which are aggregates of many small crystals. **33** and **34** show a number of glauconite pellets in a coarse sandstone. The glauconite is easily identifiable in the photograph taken in plane-polarised light by its green or brownish-green colour. The grain in upper centre part of the field of view incorporates a number of silt-sized quartz grains. Glauconite has moderate birefringence, but as the picture taken with crossed polars shows, interference colours are masked by the natural colour of the mineral. The remainder of the sediment consists of monocrystalline quartz grains and in the lower right portion of the field of view, a sedimentary rock fragment. The cement, which is showing high order interference colours, is calcite.

35 shows a sandstone rich in glauconite and containing subrounded quartz grains (low relief) and carbonate grains and cement (high relief). Note that many of the bright green glauconite pellets have brown margins. These are limonite and result from the oxidation of the ferrous iron in glauconite.

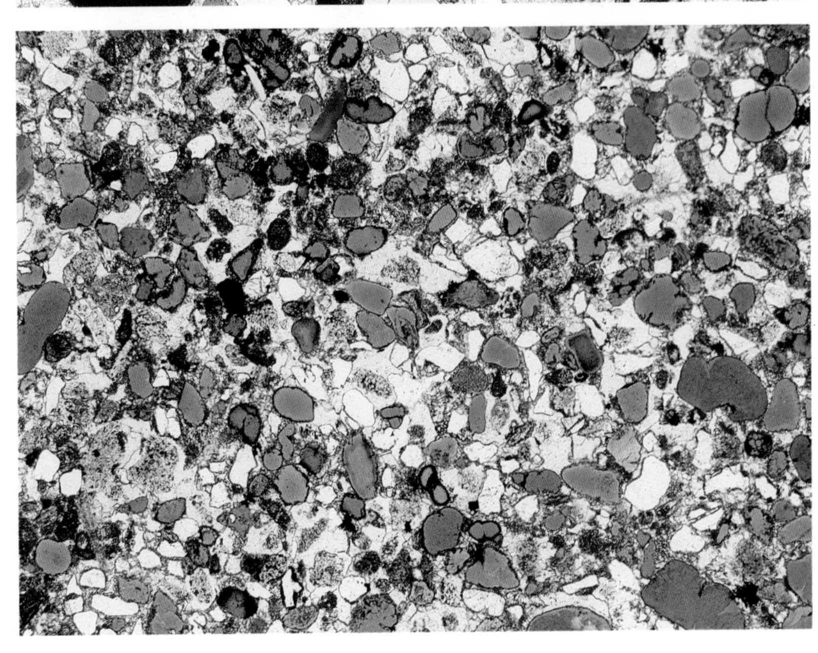

33 and 34: Lower Greensand, Lower Cretaceous, Folkestone, England; magnification × 22; 33 PPL, 34 XPL.
35: Lower Cretaceous, Co. Antrim, Northern Ireland; magnification × 22, PPL.
Glauconite is shown also in **214** *and* **215**.

36, 37, 38

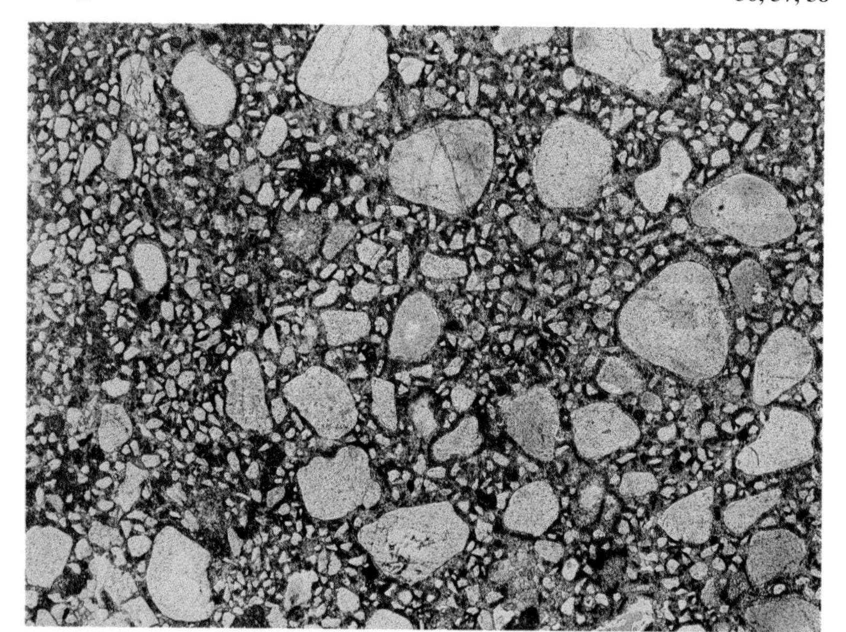

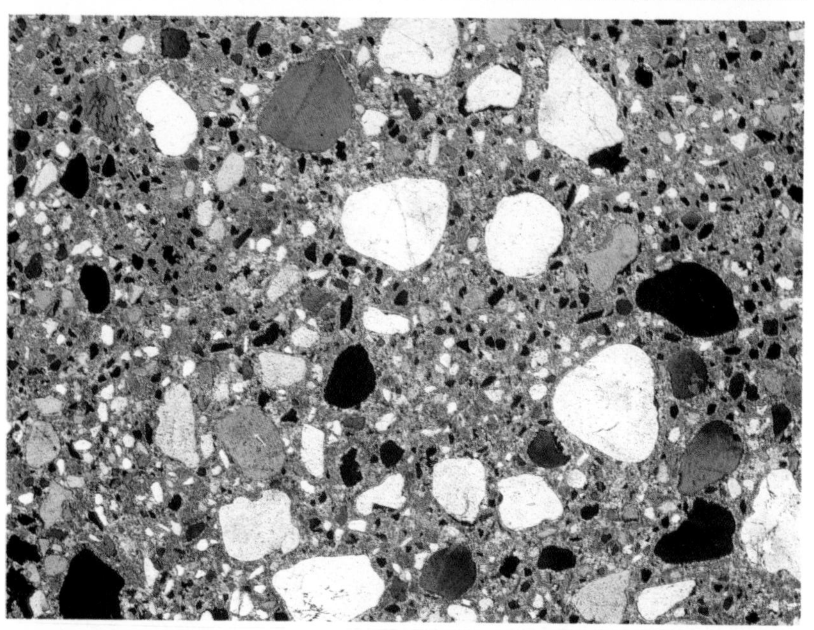

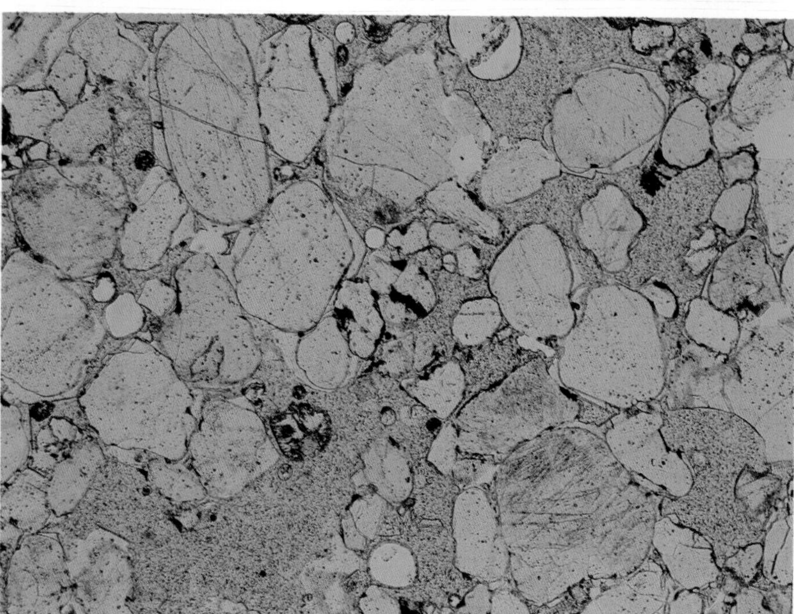

Sandstones – Matrix and cement

On deposition, many sandstones contain little sediment matrix between the component grains. Some terrigenous mud may be deposited with the grains and those sediments with more than 15% clay matrix are classified as greywackes (**62–67**). A few sandstones have a matrix of carbonate mud. **36** and **37** show a sediment containing large, rounded quartz grains together with smaller, subangular to subrounded grains in a fine-grained matrix having high relief. In the XPL photograph, high-order interference colours, characteristic of calcite, can be seen. This sample is a sandstone with a carbonate mud matrix, which was probably deposited at the same time as the grains, rather than being introduced later as a cement.

Cementation is the principal process leading to porosity reduction in sandstones, the most common cements being quartz, calcite and clay minerals. Clay mineral coatings on component grain surfaces are important in the diagenesis of sediments, in that they may inhibit the growth of pore-filling quartz or calcite cements. Such textures require the use of the electron microscope for detailed study.

38 and **39** show a highly porous sandstone with rounded quartz grains. The speckled areas which appear black in the XPL photograph are pores filled with the mounting medium. Although comprising a loose fabric of grains, the sandstone is well-cemented by secondary (authigenic) quartz in the form of overgrowths on the detrital grains. The surfaces of the original grains are picked out by a thin red-brown rim of iron oxide. Since both the overgrowth and the detrital cores of each grain show uniform interference colours, it is clear that the overgrowths grew in optical continuity with the grains on which they nucleated. Note that where overgrowths are well-developed, the overall shape of the grains has changed from rounded to subhedral. A good example of euhedral crystal terminations can be seen near the top of the photograph on the right-hand side.

Calcite cements in sandstones are usually fairly coarse-grained (sparite p. 34). Occasionally they are so coarse that one cement crystal envelopes many detrital grains, resulting in a *poikilitic* texture. **40** and **41** show a sandstone in which the detrital grains are subangular to subrounded quartz. The cement is calcite of such a grain size that there are only a few crystals in the field of view shown. Individual cement crystals can be distinguished in the XPL photograph by their slightly different interference colours (high-order grey and pink).

Sandstones – Matrix and cement

(continued)

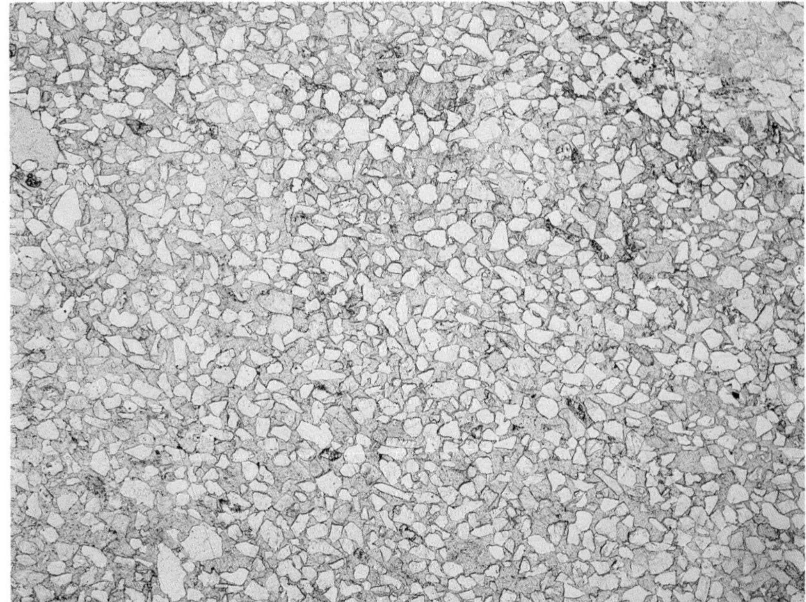

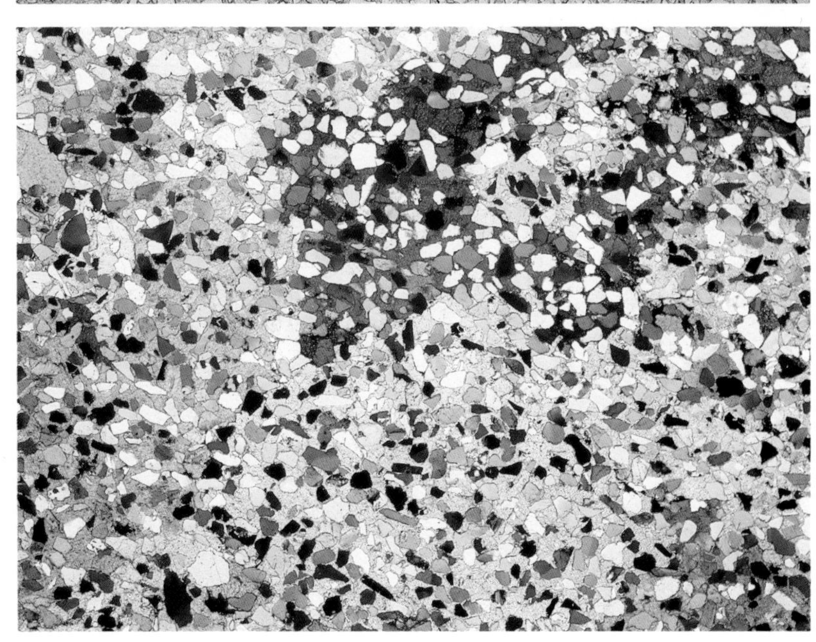

***36** and **37**: locality and age unknown; magnification × 16; **36** PPL, **37** XPL.*
***38** and **39**: Penrith Sandstone, Permian, Penrith, Cumbria, England; magnification × 27; **38** PPL, **39** XPL.*
***40** and **41**: Middle Jurassic, Bearreraig Bay, Isle of Skye, Scotland; magnification × 20; **40** PPL, **41** XPL.*

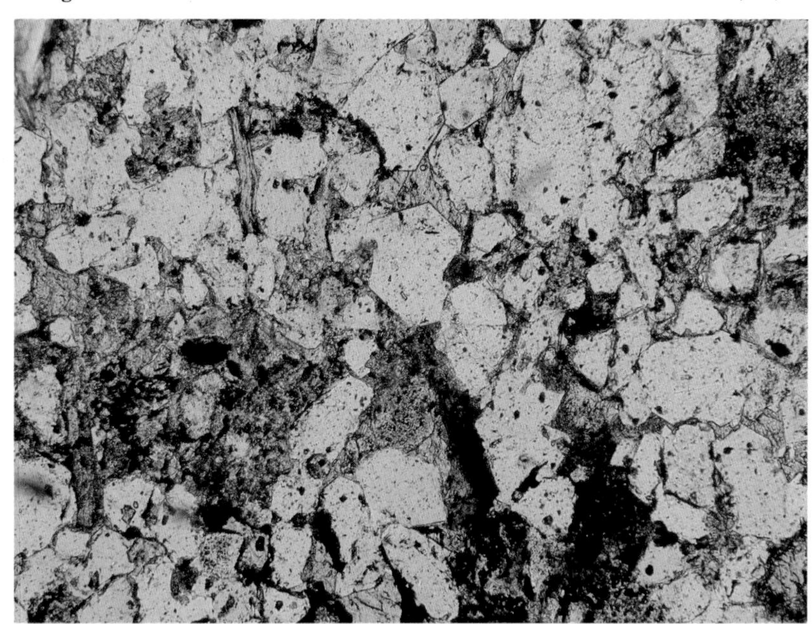

Cements

(continued)

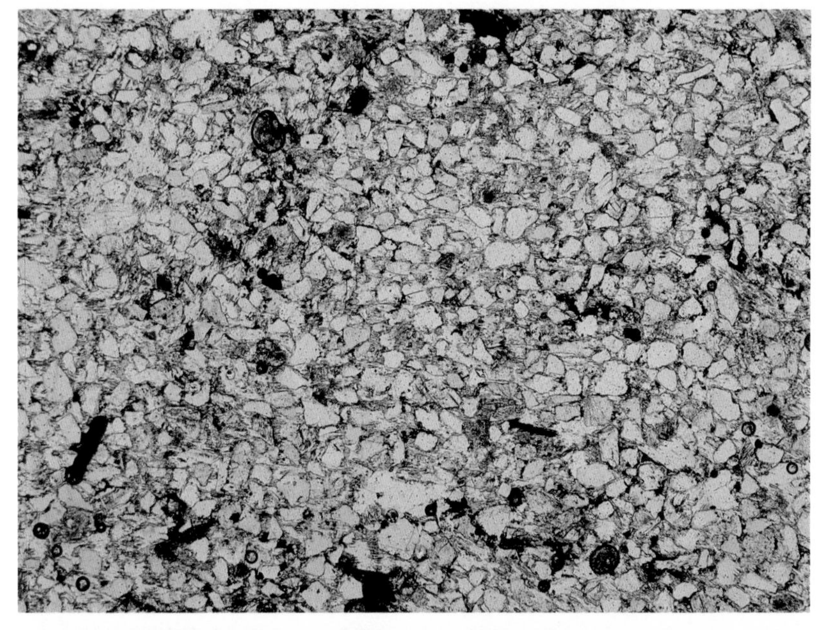

42 shows a high magnification view of a fine sandstone which contains both quartz and calcite cements. The quartz cement is in the form of overgrowths on detrital grains. Evidence for this is the euhedral terminations seen on some grains (good examples can be seen just above the centre of the photograph). Unlike the quartz in the sample in **38** and **39**, the shape of the original detrital particles is not visible where overgrowths are present. Calcite cement postdates the quartz overgrowths and infills pores. The thin section has been stained with Alizarin Red S and potassium ferricyanide (see p. 34) and the calcite is a very pale mauve colour because it contains some iron.

43 and **44** show a fine sandstone which is cemented by gypsum. Gypsum has approximately the same birefringence as quartz and so it does not show up well in the XPL photograph. In the PPL view, the higher relief of the gypsum and its cleavage help to distinguish it from quartz. Some of the gypsum cement crystals enclose several detrital grains. One showing pale grey interference colours in the XPL view occupies the upper left part of the photograph.

42: Middle Jurassic, Yorkshire, England; magnification × 72, PPL.
43 and 44: Cretaceous, Tunisia; magnification × 24; ***43*** *PPL,* ***44*** *XPL.*

Cements

(continued)

45 and **46** show a quartz sandstone at high magnification. Note the mica flake in the centre of the photograph. In the field of view shown many of the intergranular pores are unfilled (e.g. lower left) and are thus black in the XPL view. However, the quartz grains and mica flake in the centre of the view are surrounded by numerous small crystals with low relief and showing first order grey interference colours. These are clay minerals in the form of a cement. Usually an electron microscope is needed to demonstrate the shapes of clay mineral crystals and techniques such as X-ray diffraction to determine the exact identity of the minerals. In the example shown, the crystals are large enough for the typical low birefringence of kaolinite to be seen, together with the 'book' texture which develops as a result of the characteristic form of a series of stacked platy crystals. This is best seen immediately above and to the right of the mica flake.

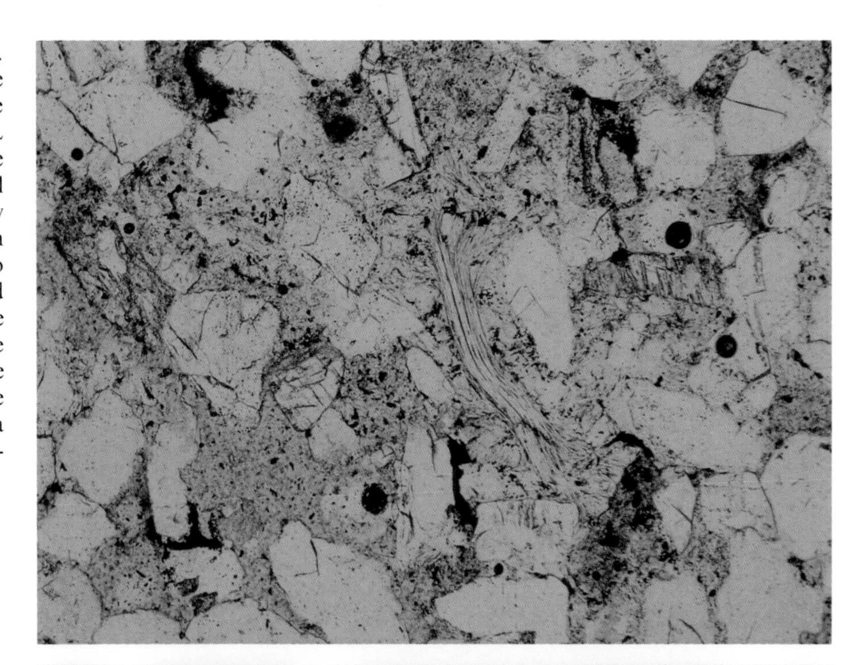

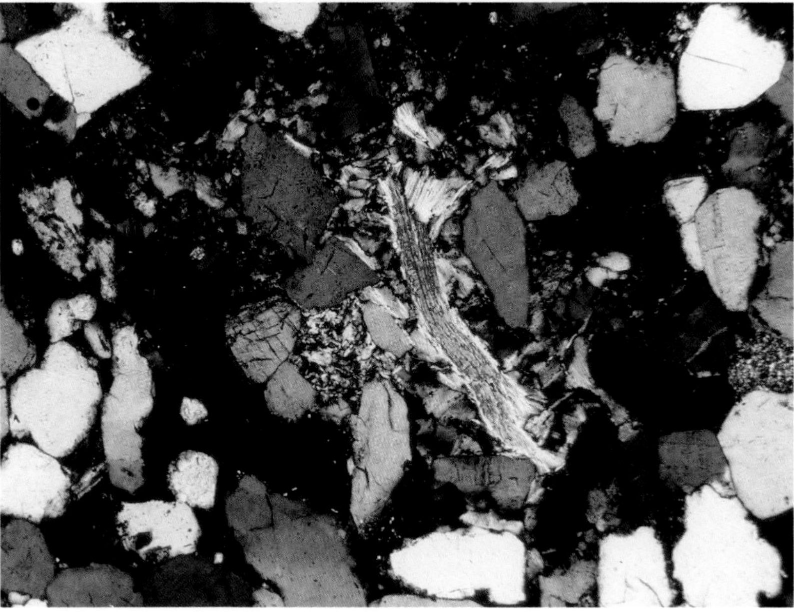

***45** and **46** Lower Carboniferous, Fifeshire, Scotland: magnification × 90; **45** PPL, **46** XPL.*

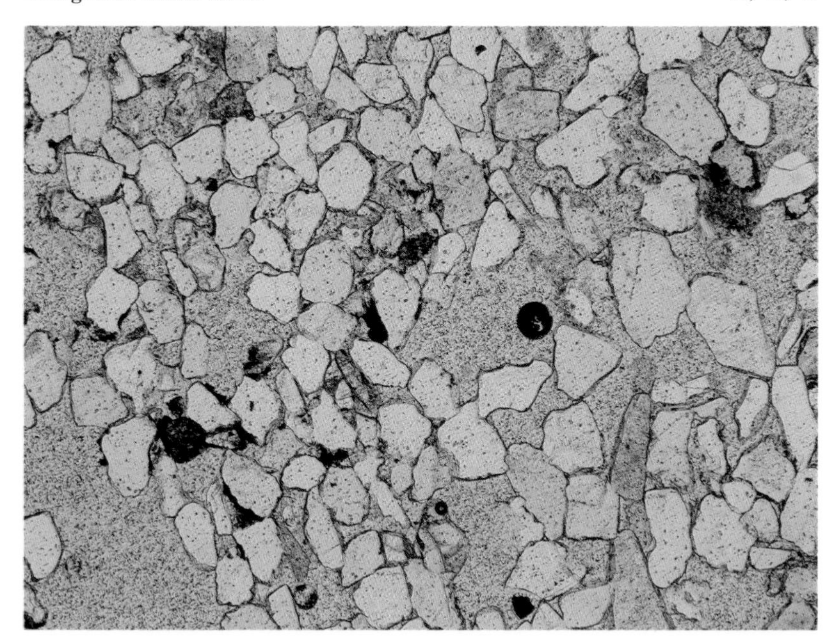

Compaction – Pressure-solution

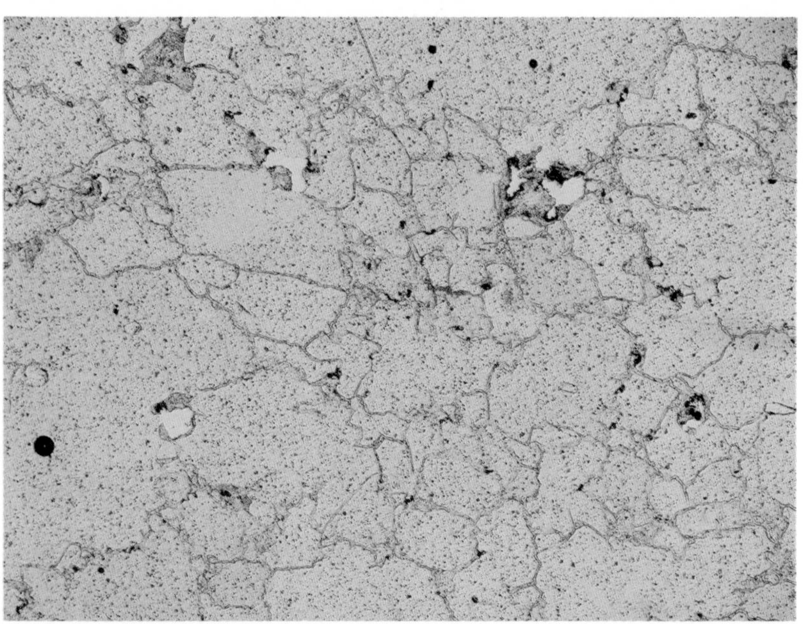

Sandstones which are not cemented early in diagenesis usually show signs of compaction. Since most of the grains in sandstone are rigid, there is usually little evidence for grain fracture and breakage (cf. limestone compaction, p. 58). Thus, apart from grain repacking during early compaction, the most important process of compaction is *pressure-solution*. This is the process whereby a sediment under load is subject to selective solution.

47 shows a sandstone with high intergranular porosity (the pores, now filled with the mounting medium, are the grey areas). Most of the quartz grains are coated with a thin brown rim of hematite cement. At many of the contacts between grains, one quartz grain has undergone solution leading to the penetration of one grain by another (concavo-convex contacts). Good examples can be seen in the upper left part of the photograph. This is the first stage of pressure-solution.

Where pressure-solution is more intense, the contacts between grains become sutured. **48** and **49** show a sandstone in which grain contacts are irregular and wavy because of pressure-solution. Silica dissolved during the process may be precipitated as cement away from grain contacts, leading to the destruction (occlusion) of porosity. As can be seen, the result is a texture in which the original grain boundaries can no longer be identified. The sample illustrated is particularly unusual in that a thin zone of clay or mica separates the sutured quartz grains. It has a higher relief than the quartz and is clearly visible in the photograph taken with PPL. This thin zone of material together with the sutured contacts enables the quartz grains to move slightly relative to their neighbours. This property imparts flexibility to the sandstone, demonstrable in hand specimens. Sandstones of this type are known as *flexible sandstones* or *itacolumites* and are extremely rare.

47: New Red Sandstone, Triassic, Cheshire, England; magnification × 43, PPL.
***48** and **49**: Itacolumite, Brazil; magnification × 31; **48** PPL, **49** XPL.*

Grain solution and replacement

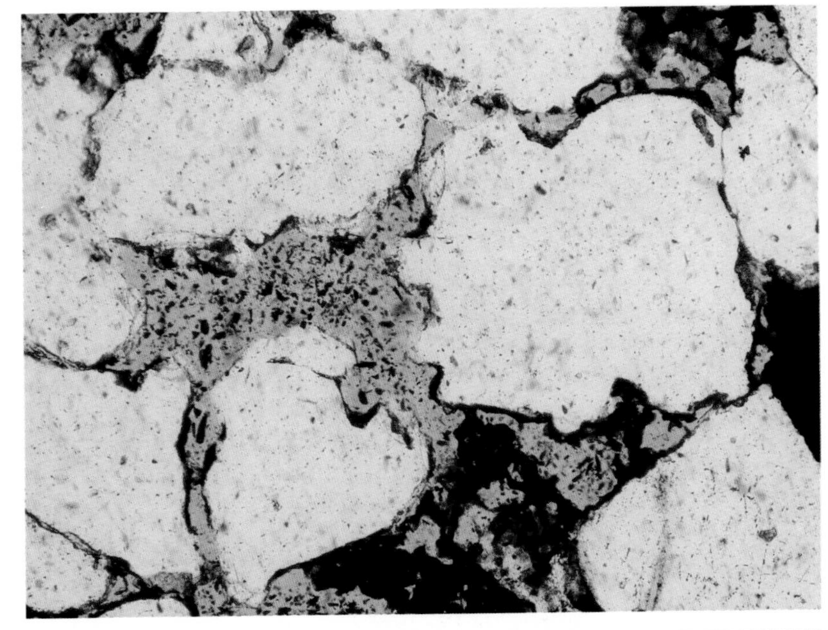

50 shows a porous sandstone. In this example the mounting medium has been impregnated with a dye so that the pores appear mauve-coloured. Note that the margins of some of the quartz grains are embayed. This has occurred as a result of corrosion of the quartz during diagenesis and has led to enhancement of the porosity. The common porosity types in sediments are illustrated in Fig. F (see p. 65) and in **151–160**, with examples from limestones. Many of the terms are equally applicable to sandstones.

51 and **52** show a sandstone cemented by a few large calcite crystals. Note the typical high-order interference colours displayed by the calcite, seen in the XPL photograph. The detrital grains, including both monocrystalline and polycrystalline quartz, are coated with a thin brown rim of iron oxide. The texture of the rock is unusual, in that it is apparently not grain-supported, being about 30% quartz grains and 70% calcite. Replacement of original detrital grains by calcite is partly responsible for this appearance, areas of calcite outlined by iron oxide being interpreted as the original grains. A good example can be seen in the centre.

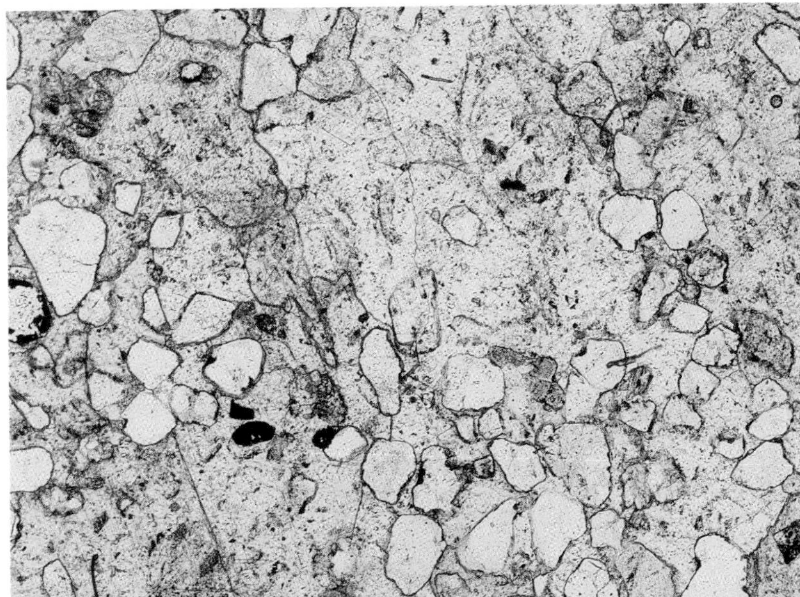

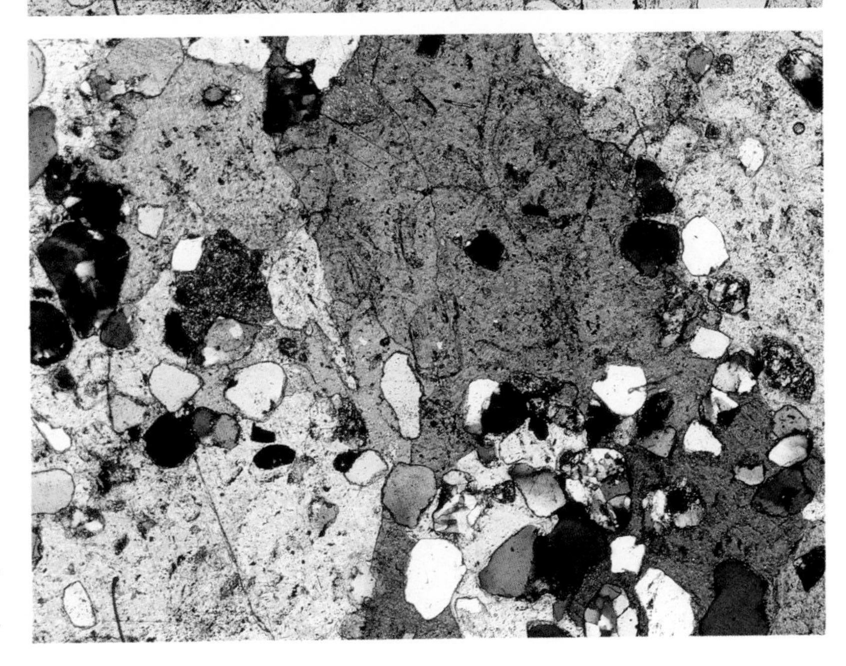

50: Saltwick Formation, Middle Jurassic, Eskdale, Yorkshire, England; magnification × 132, PPL.
51 and 52: New Red Sandstone, Triassic, Brixham, Devon, England; magnification × 43; ***51*** *PPL,* ***52*** *XPL.*

Sandstone classification

Modern sandstone classifications require the estimation of the proportions of the principal grain types and thus thin section study is required.

Of the different sandstone classifications proposed, we present a widely-used example, that of Folk (1974). Figure B shows the classification of those rocks containing less than 15% fine-grained matrix in terms of the three principal components; quartz, feldspar (plus granite and gneiss fragments) and other rock fragments. Those sandstones containing more than 15% fine-grained matrix are the greywackes, and are subdivided according to Fig. C. We refer readers to Pettijohn (1975) for details of other sandstone classifications and for the classification of conglomerates and mudrocks, where studies using the petrographic microscope are less important.

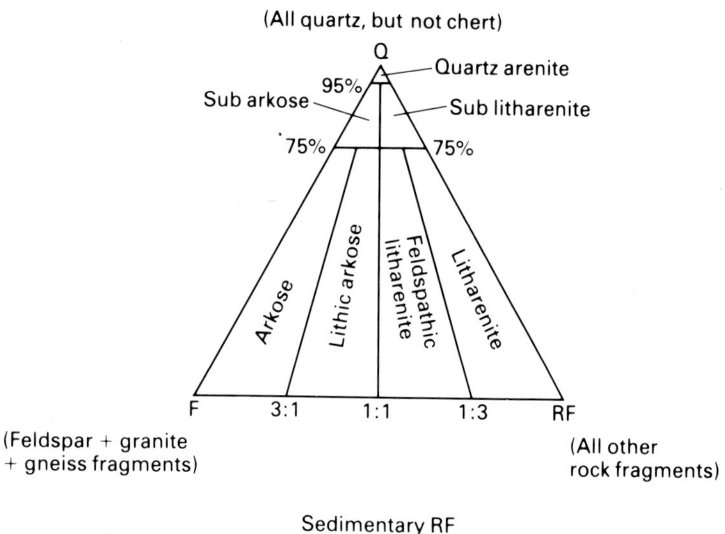

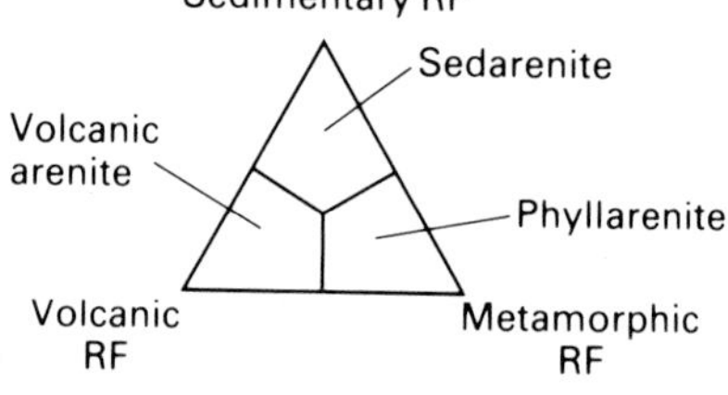

Fig. B Classification of sandstones. The upper triangle shows a sandstone classification for sediments with less than 15% fine-grained matrix. Classification involves the removal of matrix, cement, micas etc, and recalculation of components to 100%. The lower triangle shows how litharenites may be further classified. (From Folk, 1974)

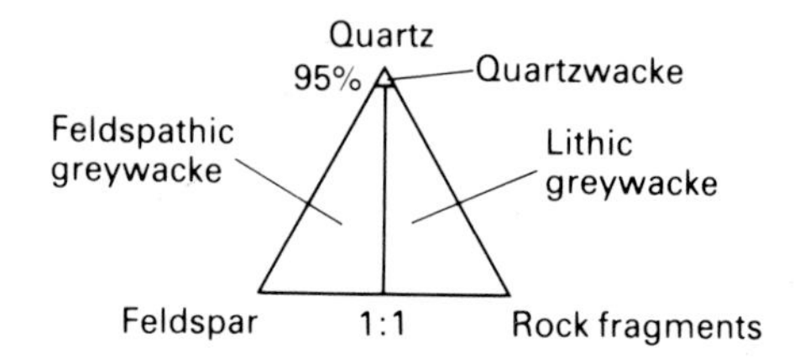

Fig. C Classification of sandstone with more than 15% fine-grained matrix (greywackes)

Sediment maturity

There are two types of sediment maturity – mineralogical and textural. Mineralogically mature sediments are those containing a high proportion of the most chemically stable and most physically resistant minerals such as quartz, chert and ultrastable heavy minerals, such as zircon and tourmaline. Mineralogically immature sediments contain the less stable grains, such as feldspars, and those rock fragments not consisting principally of quartz.

The textural maturity of a sediment depends on the content of fine-grained material, the sorting and the roundness of the grains. A scale of textural maturity proposed by Folk (1951) is presented below.

Immature stage – Sediment contains >5% clay matrix. Grains poorly-sorted and not well-rounded.

Submature stage – sediment contains <5% clay matrix. Grains poorly-sorted and not well-rounded.

Mature stage – sediment contains little or no clay. Grains well-sorted, but not well-rounded.

Supermature stage – sediment contains no clay. Grains well-sorted and well-rounded.

Diagrams illustrating visual estimation of sorting sediments using thin sections are shown in Fig. D.

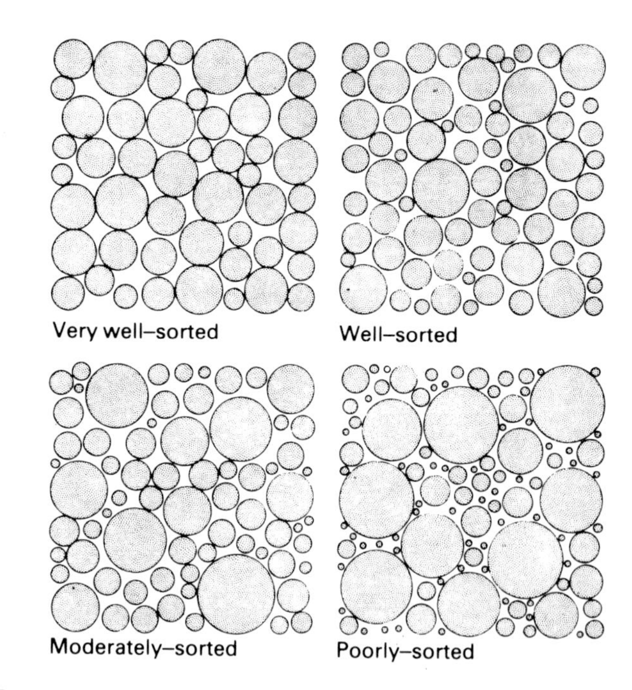

Fig. D Sorting – thin sections (after Pettijohn et al., *1973)*

Quartz arenite, arkose

53 shows a sandstone which consists almost entirely of quartz and is thus classified as a quartz arenite. Such sandstones were called quartzites in older classifications, although it is perhaps better to restrict the term quartzite to metamorphic rocks. Since they contain more than 95% quartz, quartz arenties are always mineralogically mature. The example shown is texturally submature to mature, lacking clay and being reasonably well-sorted. Rounding of the grains is difficult to assess because the effects of compaction and cementation have obscured the shape of the original grains.

54 and **55** show a sediment in which more than 50% of the grains are feldspar, easily identifiable in PPL by the brown colour resulting from their alteration (see p. 7) and in crossed polars by the remnants of multiple twinning in many grains. Most of the other grains in the sediment are clear quartz, so the sample is an arkose. A sediment with such a high proportion of relatively unstable feldspar grains is mineralogically immature. The matrix contains abundant opaque iron oxide.

53: Millstone Grit, Upper Carboniferous, Craig-y-Dinas, South Wales; magnification × 27, XPL.
***54** and **55**: Torridonian, Precambrian, Scotland; magnification × 20; **54** PPL, **55** XPL.*

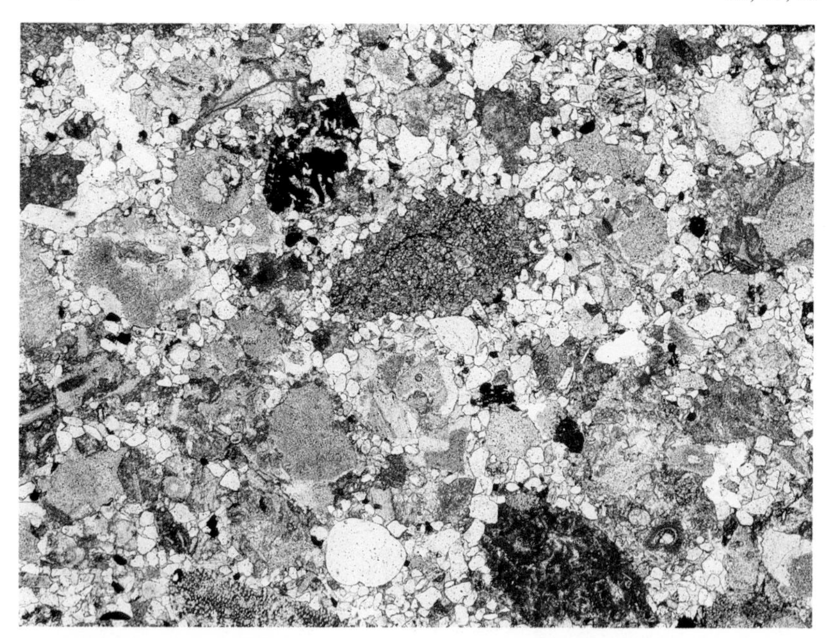

Litharenites

Litharenites are sandstones with less than 95% quartz and more rock fragments than feldspar. They may be classified according to whether the rock fragments are predominantly sedimentary, volcanic or metamorphic (Fig. B p. 24).

56 and **57** show a sedarenite, in which the fragments are from carbonate rocks. The fine-grained fragment just above the centre is from dolomite rock. Examples of limestone fragments can be seen in the lower right-hand quadrant. The sediment also contains monocrystalline quartz and echinoderm plates. The latter are the speckled grains with uniform interference colours (see p. 44). In this example the echinoderms are reworked from an older limestone and are not fragments of fossils living at the time of final deposition of the sediment. Thus they are classified as sedimentary rock fragments rather than as fossil material.

58 and **59** show a mineralogically immature sediment consisting mainly of igneous rock fragments cemented by pale brown chlorite. The clear areas in the view taken with PPL show high-order interference colours under crossed polars and are carbonate. A variety of grain types is present and all show some signs of alteration. Many of the rock fragments contain partially-altered phenocrysts of plagioclase in a groundmass of plagioclase laths and another mineral too fine-grained to determine, but which may be chlorite. Individual plagioclase crystals are also present and vary from euhedral laths to subhedral grains. The porphyritic texture of the igneous rocks suggests a volcanic source rock and thus the sediment is a volcanic arenite. Such an immature sediment would be very close to its source rock and it is, perhaps, a locally-reworked pyroclastic rock.

60 and **61** show a sediment which is more than 75% quartz. The remaining grains are rock fragments and hence the sediment may be classified as a sublitharenite. The rock fragments are of fine-grained sedimentary and metasedimentary rocks.

Litharenites

(continued)

56 *and* ***57****: Brockram, Permian, Appleby, Cumbria, England; magnification × 16;* ***56*** *PPL,* ***57*** *XPL.*
58 *and* ***59****: Ordovician, Builth Wells, Powys, Wales; magnification × 12;* ***58*** *PPL,* ***59*** *XPL.*
60 *and* ***61****: Coal Measures, Lancashire, England; magnification × 14;* ***60*** *PPL,* ***61*** *XPL.*

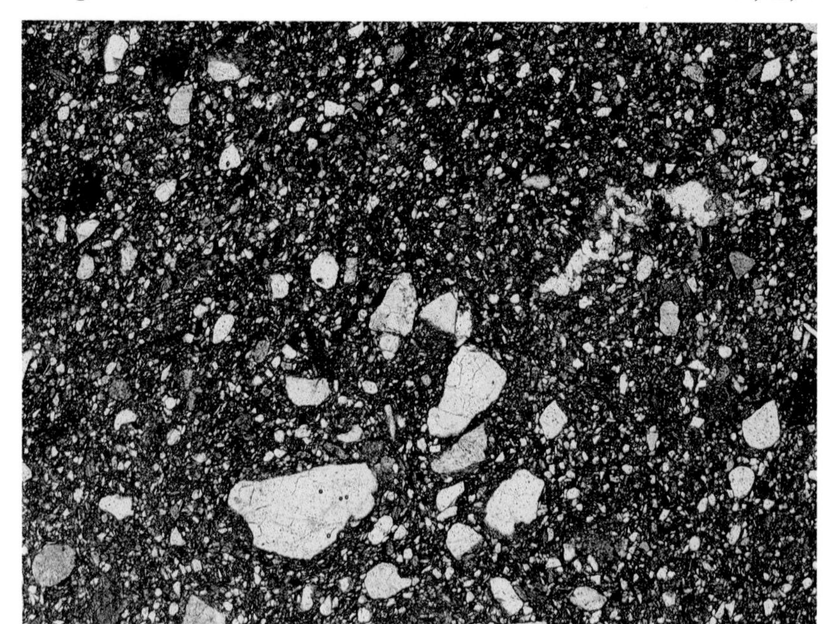

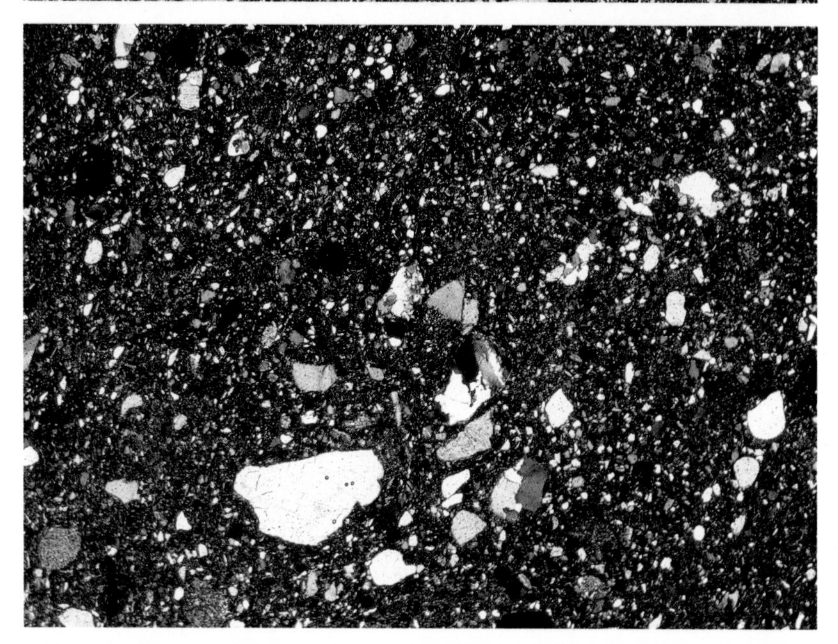

Greywackes

Greywackes are those sandstones containing more than 15% fine-grained matrix. Their classification is shown in Fig. C (see p. 24).

62 and **63** show a typical greywacke, being poorly-sorted and containing abundant fine-grained matrix (almost opaque in the view taken with plane-polarised light). The fragments are predominantly monocrystalline and polycrystalline quartz grains, but a small percentage of rock fragments (cloudy particles of fine-grained material) make this a lithic greywacke.

64 and **65** illustrate a sediment with about 15% matrix and containing quartz and abundant feldspar grains. Feldspars include plagioclase with multiple twinning and perthitic alkali feldspar. The sediment is thus classified as a feldspathic greywacke.

66 and **67** show a greywacke in which quartz, feldspar and rock fragments are visible. Quartz grains are clear in the PPL view, whereas the feldspars are brownish owing to alteration. The XPL view shows that some feldspar grains are multiple-twinned plagioclase whereas others are microcline showing typical cross-hatch twinning (e.g. right of field of view; about half way up). The grain in the centre of the field of view is an igneous rock fragment consisting of plagioclase and amphibole. The amphibole can be recognized by its green absorption colour and its two cleavages at 120°. Smaller fine-grained rock fragments and individual ferro-magnesian mineral grains are also present.

Greywackes

(continued)

***62** and **63**: Ashgillian, Dyfed, Wales; magnification × 16; **62** PPL; **63** XPL.*
***64** and **65**: Locality and age unknown; magnification × 16; **64** PPL, **65** XPL.*
***66** and **67**: Silurian, Peebleshire, Scotland; magnification × 43; **66** PPL, **67** XPL.*

Siltstones

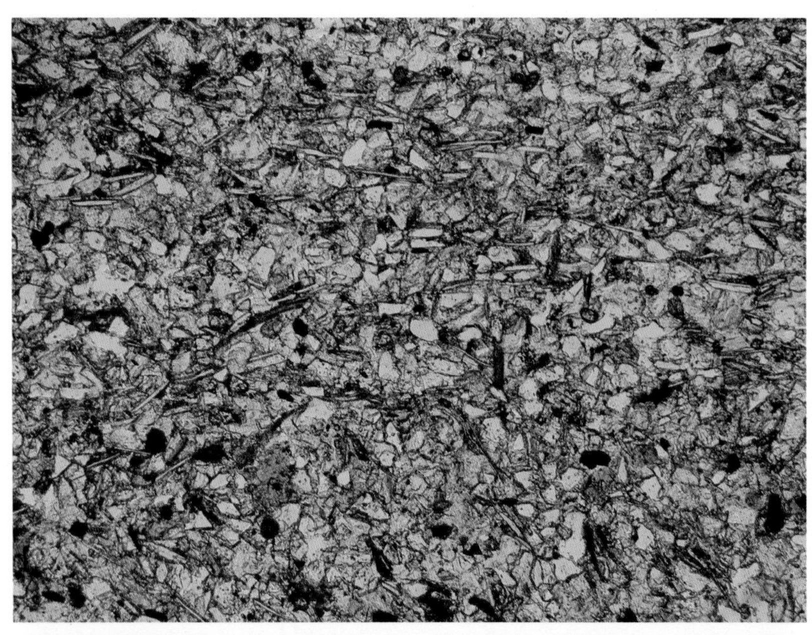

Siltstones are those terrigenous sediments in which the majority of the grains are between $\frac{1}{16}$ and $\frac{1}{256}$ mm in diameter (Table 1, see p. 3). **68** and **69** show a coarse siltstone (notice that the magnification is higher than that of most of the previous photographs) with abundant quartz grains and thin mica flakes. The micas include both muscovite (colourless) and biotite (yellow or brown). Muscovite is more abundant and shows second-order interference colours, seen in XPL. The sediment is cemented by calcite, showing high relief in PPL and high-order interference colours with polars crossed.

68 and 69: Age and locality unknown; magnification × 72; 68 PPL, 69 XPL.

Siltstones

(continued)

Many siltstones show small-scale sedimentary structures. **70** shows a laminated siltstone, the laminae being defined by changes in grain size. The dark layers, seen near the base of the photograph, are composed almost entirely of clay-sized material, whereas the band just above the centre is composed of clear fine sand-sized quartz. The photograph also shows graded layers (below and above the coarse band). The fining-upwards grading is shown by a decrease in the clear quartz material and an increase in the dark-coloured clay.

71 shows a siltstone in which ripple cross-lamination can be seen, indicating a flow from right to left, as seen in the picture. The ripple structure is picked out by the alternation of dark clay-rich and pale clay-poor laminae.

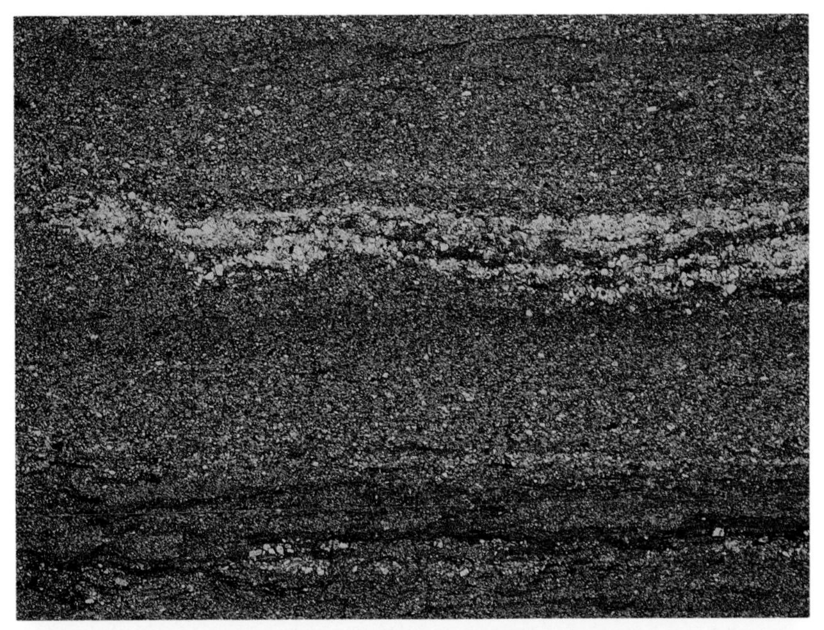

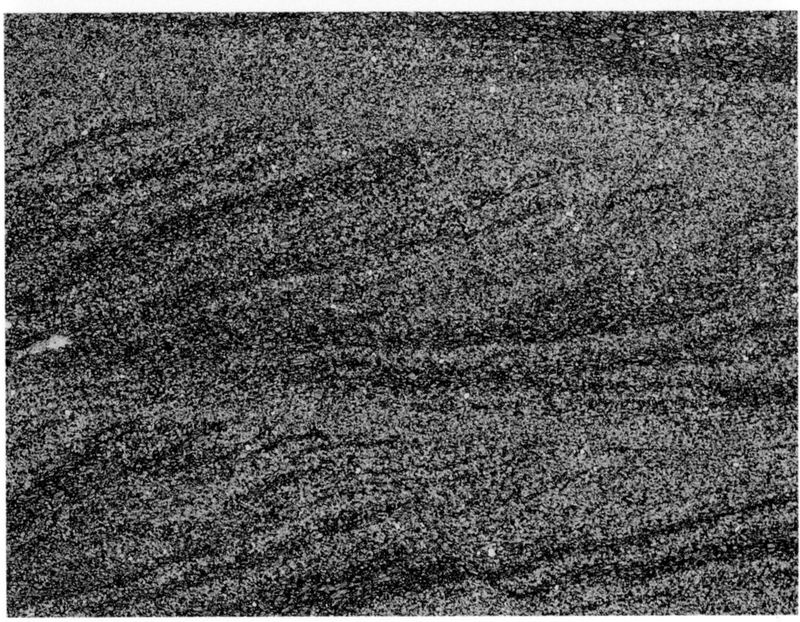

70: Coal Measures, Upper Carboniferous, Lancashire, England; magnification × 15, PPL.
71: Ashgillian, Llangranog, Dyfed, Wales; magnification × 9, PPL.

Part 2

Carbonate rocks

Introduction

Unlike terrigenous sediments, carbonate rocks comprise material formed mostly at or near the site of final accumulation of the sediment. Much of the material is produced by biological processes. Two carbonate minerals are common in older limestones – *calcite*, $CaCO_3$, and *dolomite*, $CaMg(CO_3)_2$. In recent shallow marine carbonate sediments the mineral *aragonite*, also $CaCO_3$, is abundant. However, it is metastable under the normal conditions prevailing in sediments and is usually dissolved once a limestone is exposed to circulating meteoric waters. Alternatively it may invert directly to calcite. Dolomite is normally a secondary mineral replacing calcium carbonate, although it may form in sediments very soon after their deposition. Both calcite and dolomite may contain some ferrous iron, in which case the prefix *ferroan* is used before the mineral name.

The optical properties of calcite and dolomite are similar and therefore they can be difficult to distinguish optically. Simple chemical staining techniques are often employed by carbonate sedimentologists to distinguish calcite from dolomite and to distinguish ferroan from non-ferroan minerals.

The dye *Alizarin Red S* is used to differentiate calcite and dolomite, whereas *potassium ferricyanide* is used to differentiate ferroan and non-ferroan minerals. The dyes are dissolved in a weak acid solution. This also helps to distinguish dolomite from calcite, as dolomite does not react with cold dilute acid whereas calcite does, producing a contrast in relief between the two minerals. Results of the etching and staining process are shown in Table 2. Details of the procedure are given in Appendix 2.

Table 2. Etching and staining characteristics of carbonate minerals

Mineral	Effect of etching	Stain colour with Alizarin Red S	Stain colour with potassium ferricyanide	Combined result
Calcite (non-ferroan)	Considerable (relief reduced)	Pink to red-brown	None	Pink to red-brown
Calcite (ferroan)	Considerable (relief reduced)	Pink to red-brown	Pale to deep blue depending on iron content	Mauve to blue
Dolomite (non-ferroan)	Negligible (relief maintained)	None	None	Colourless
Dolomite (ferroan)	Negligible (relief maintained)	None	Very pale blue	Very pale blue (appears turquoise or greenish in thin section)

The intensity of the stain colour is partly related to the intensity of the etching with acid. Fine-grained crystal fabrics with many crystal boundaries etch more rapidly and thus show deeper stain colours than coarse crystal fabrics with few crystal boundaries.

Stain colours are particularly well-illustrated in **100**, **124**, **131**, **161** and **165**.

Other stains have been used to distinguish between aragonite and calcite and to identify magnesian calcites: details are given in books on techniques in sedimentary petrology such as that of Carver (1971).

Carbonate rocks may also be examined using acetate peels. These record an impression of an etched rock surface (which also may be stained) on a thin sheet of acetate film. Acetate peels have the advantage of being cheap and easy to make, but because they are isotropic, minerals cannot be identified by optical properties, such as relief and birefringence. Details of the procedure for making acetate peels are given in Appendix 3.

Components

The three most important components of carbonate rocks are *allochemical components*, *microcrystalline calcite*, and *sparry calcite*.

1. Allochemical components or *allochems*, are organized aggregates of carbonate sediment which have formed within the basin of deposition. They include ooids, bioclasts, peloids, intraclasts and oncoids and are considered in detail in the following section (**72** to **120**).
2. Microcrystalline calcite or *micrite* is carbonate sediment in the form of grains less than 5 μm in diameter. Much of it forms in the basin of deposition, either as a precipitate from seawater or from the disintegration of the hard parts of organisms, such as green algae. The term 'carbonate mud' is also used for this fine sediment, although strictly mud includes material of clay- and silt-size (up to 62 μm). Micrite is illustrated in **84**, **89**, **111** and **157**.
3. Sparry calcite, *sparite* or *spar* refers to crystals of 5 μm or more in diameter. Much of it is coarse, with crystals commonly up to 1 mm in size. It is usually a pore-filling cement and thus may form in a rock a long time after deposition of the original allochems and micrite. Sparite is illustrated in **73**, **82**, **124** and **131**.

The classification of limestones involves the identification of allochems and estimation of the proportions of micrite and sparite (see p. 62).

Ooids

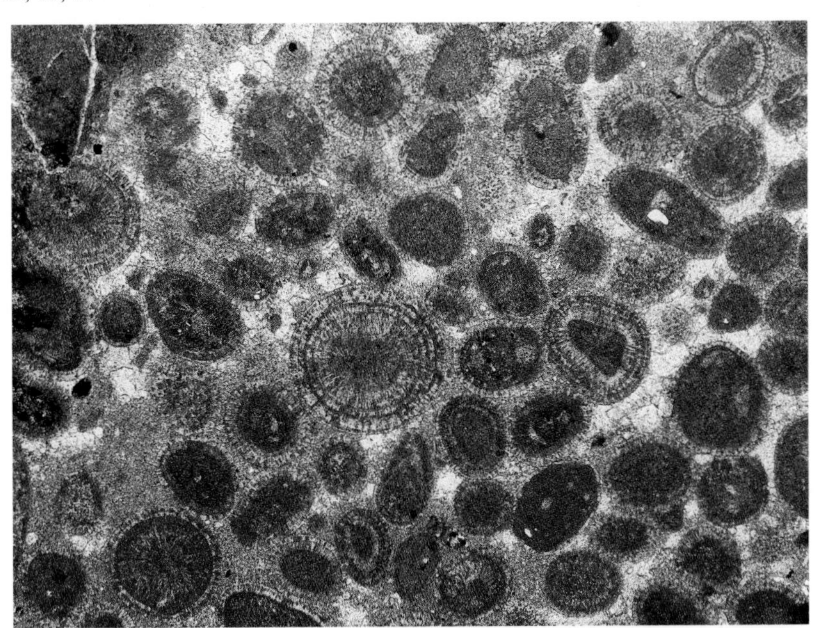

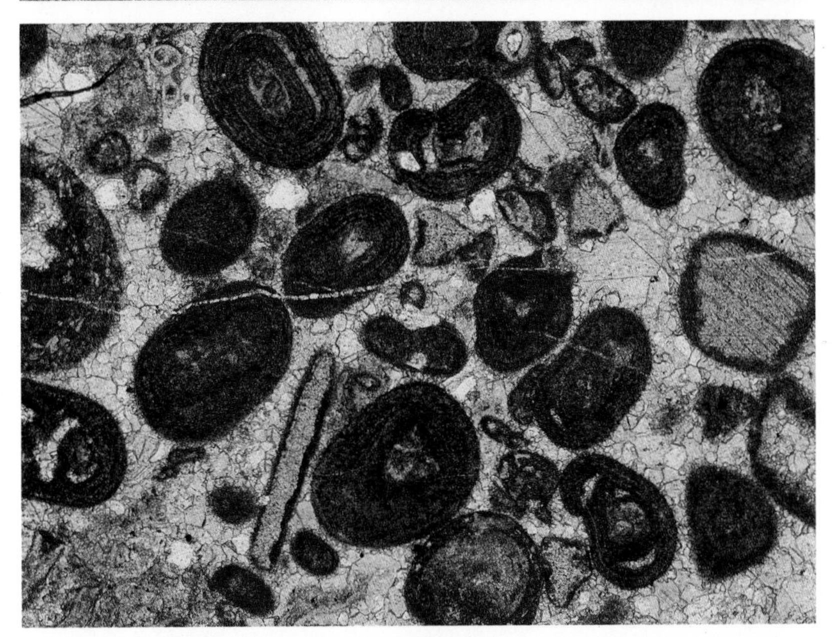

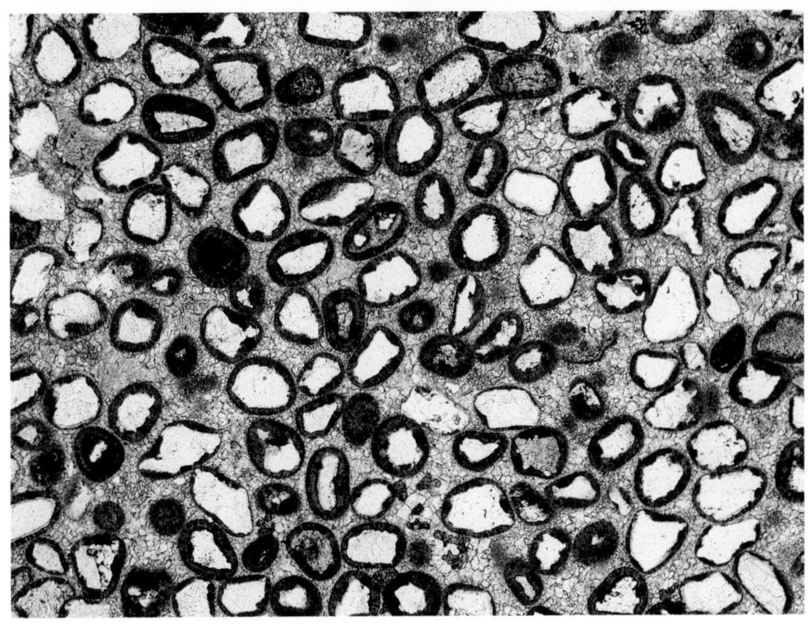

Ooids or *ooliths* are spherical or ellipsoidal grains, less than 2 mm in diameter, having regular concentric laminae developed around a nucleus. Ancient ooids often show both the concentric laminae and a radial structure. It is not always certain whether the radial structure is primary, or formed during the inversion of aragonite to calcite.

72 shows ooids with well-developed radial and concentric structures. The nuclei are micritic carbonate grains. The sample shows a range of ooids, from those with a small nucleus and thick cortex (the oolitic coating), to those with a large nucleus and a single oolitic lamina. The latter are called *superficial ooids*. The matrix between the ooids is a mixture of carbonate mud and sparry calcite cement.

73 illustrates ooids with a rather poorly-preserved concentric structure. The structure may have been partly lost by micritization (p. 54). The speckled plates with thin micrite coatings are echinoderms (an example can be seen half way up the right-hand edge). The pink-stained cement is non-ferroan sparry calcite. The unstained grains with low relief are secondary (authigenic) quartz replacing calcite.

74 shows ooids with relatively thin cortices coating detrital quartz nuclei. Note how the early ooid laminae fill in depressions on the surface of quartz grains and are absent from angular corners. The cement is pink-stained non-ferroan sparry calcite.

72: Stained thin section, Upper Jurassic, Cap Rhir, Morocco; magnification × 31, PPL.
73: Stained thin section, Hunt's Bay Oolite, Lower Carboniferous, South Wales; magnification × 43, PPL.
74: Stained thin section, Carboniferous Limestone, Llangollen, Clwyd, Wales; magnification × 27, PPL.
Ooids can also be seen in **125**, **127**, **137**, **146**, **147** *and* **155**.

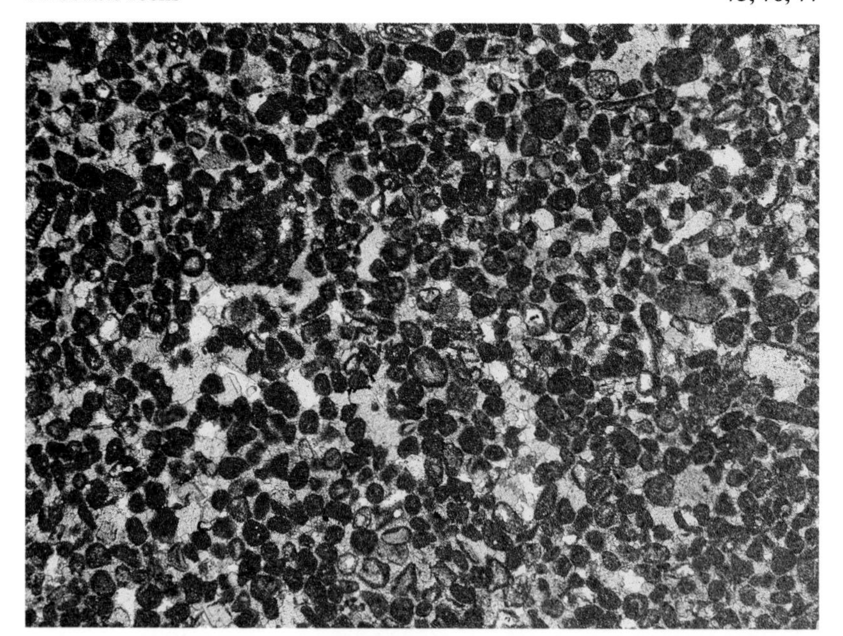

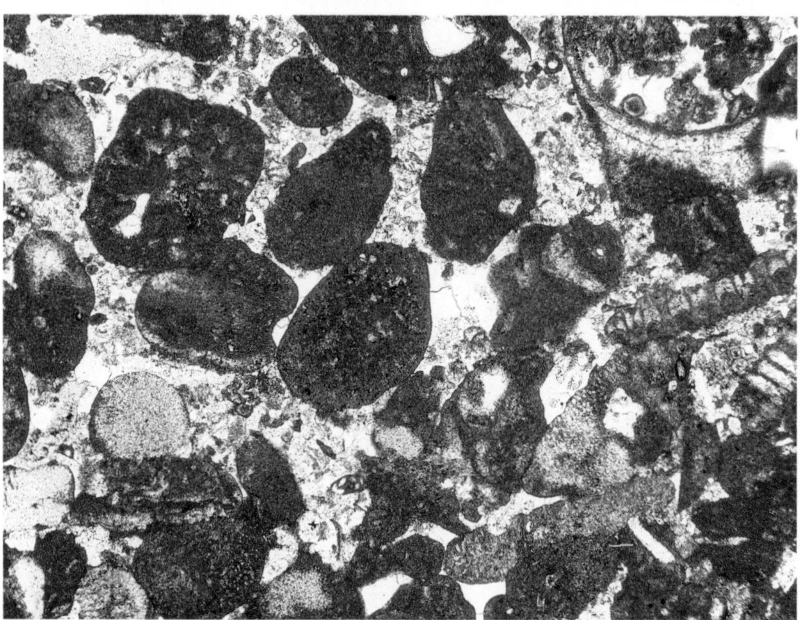

Peloids and Intraclasts

A large proportion of the allochems in limestones are grains composed partly or entirely of micrite, but having no concentric laminae in their outermost zones. Various terms have been used to classify these grains and most depend on an interpretation of their origin.

Those grains composed of micrite and lacking any recognizable internal structure are called *peloids*. **75** shows a limestone in which the allochems are mainly peloids, circular to elliptical in cross-section and averaging about 0.1 mm in diameter. Such peloids are generally interpreted as faecal in origin and are called *pellets*. The photograph shows pellets at the lower end of the size range for typical pellets, which extends up to 0.5 mm.

76 shows larger, less regular peloids, some of which have a trace of internal structure although its nature cannot be identified. In the lower part of the photograph are speckled echinoderm plates, and midway up the right hand edge are segments of the dasycladacean alga *Koninckopora* (see **113**). Both echinoderms and algae show signs of replacement by micrite around their margins (micritization, p. 54). It is probable that the peloids were formed by intense micritization of bioclasts, thus accounting for their vague relict structures.

Intraclasts are sediment which was once incorporated on the sea-floor of the basin of deposition and was later reworked to form new sediment grains. **77** shows a large grain which might be described as a 'coated bioclast'. It comprises a nucleus, which is a fragment of a brachiopod shell, surrounded by a coating of microcrystalline calcite. The coating is not laminated, so the grain cannot be called an oncoid (see p. 38); it is external to the shell and has a sharp contact with it so the coating was not formed by micritization (see p. 54). It is therefore likely that it is a fragment of locally-reworked sediment, the brachiopod shell having once been incorporated in a fine-grained sediment which was later eroded to produce intraclasts.

75: Stained thin section, Upper Jurassic, Cap Rhir, Morocco; magnification × 33, PPL.
76: Unstained thin section, Woo Dale Limestone, Lower Carboniferous, Long Dale, Derbyshire, England; magnification × 21, PPL.
77: Stained thin section, Urswick Limestone, Lower Carboniferous, Trowbarrow, Cumbria, England; magnification × 15, PPL.
*Peloids are also shown in **86**, **123**, **130**, **134**, **147**, **158** and **162**.*

Aggregate grains and lithoclasts

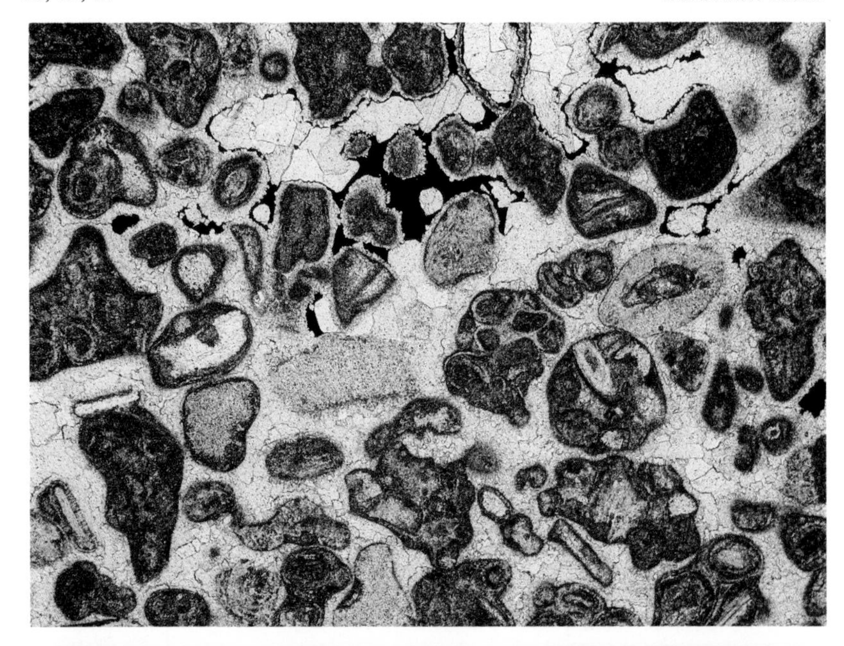

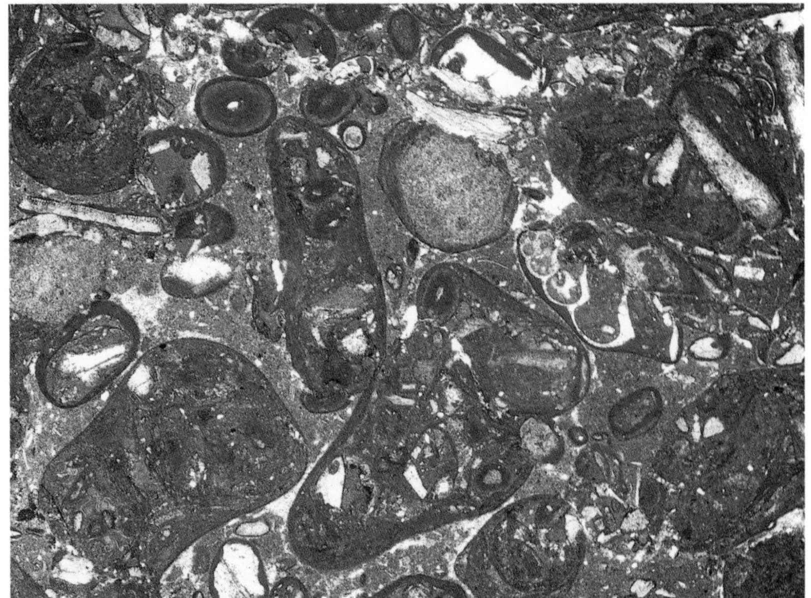

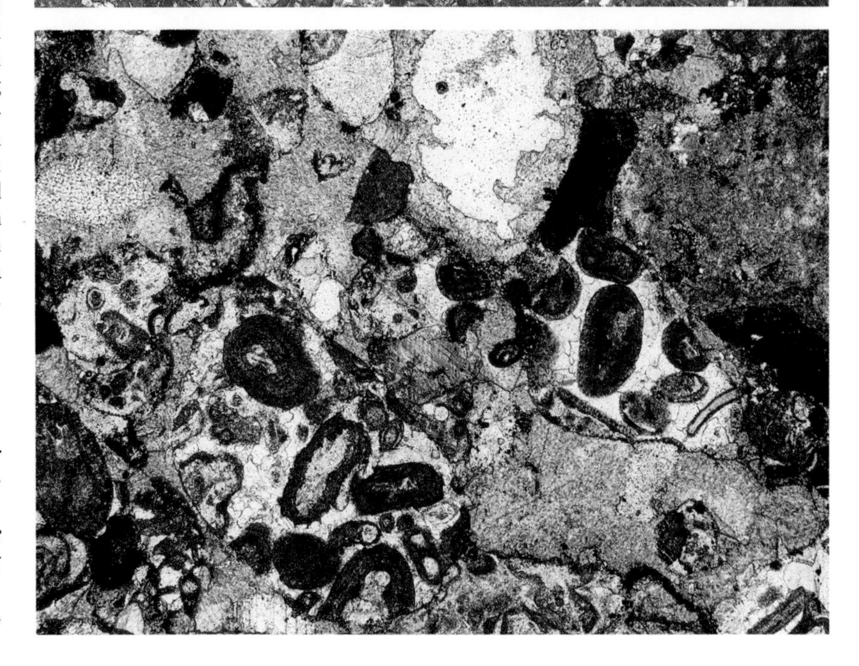

78 and **79** show *aggregate grains*. These are made up of irregular aggregates of a small number of recognizable particles cemented together by micrite or fine sparite. **78** shows the botryoidal form typical of these aggregates. The component particles include ooids (the grain right of centre) as well as peloids and a few bioclasts. These aggregates are similar to the *grapestones* of modern sedimentary environments, where particles become cemented on the sea-floor in areas of low sedimentation rate. The opaque material in the top centre is bitumen (see **160**).

79 shows large aggregate grains with a smooth, rather than botryoidal, external form. The micritic material binding the particles together completely envelopes them, and is more important volumetrically than the cementing material of the grains shown in **78**. It is unlikely that the aggregation occurred by cementation on the sea-floor, but the particles are probably reworked grains and thus could also be described as intraclasts. The matrix is micrite with a little sparite and some bioclasts.

Lithoclasts or *extraclasts* are eroded fragments of lithified sediment which have been transported and redeposited. **80** shows lithoclasts which are made up of ooids and bioclasts cemented by very pale pink-stained non-ferroan sparry calcite. Note the truncation of both particles and cement at the lithoclast margins, indicating reworking of lithified sediment. The equant sparite cement within the lithoclasts is typical of precipitation from meteoric waters (p. 55), so these fragments are of a limestone which was not cemented in either the original environment of deposition of the component particles in the lithoclasts, nor in the final environment of deposition of the clasts themselves. They are, in fact, fragments of a Carboniferous Limestone reworked during the Jurassic. The final cement is lilac-stained, coarse ferroan calcite.

78: Unstained thin section, Bee Low Limestone, Lower Carboniferous, Windy Knoll, Derbyshire, England; magnification × 27, PPL.
79: Stained thin section, Ouanamane Formation, Middle Jurassic, Ait Chehrid, Western High Atlas, Morocco; magnification × 14, PPL.
80: Stained thin section, Sutton Stone, Lower Jurassic, Ogmore-by-Sea, South Wales; magnification × 28, PPL.

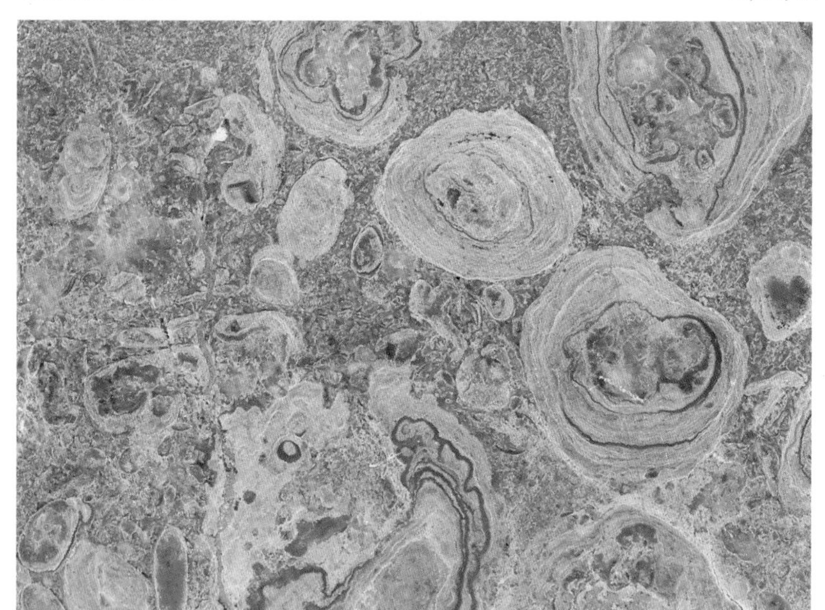

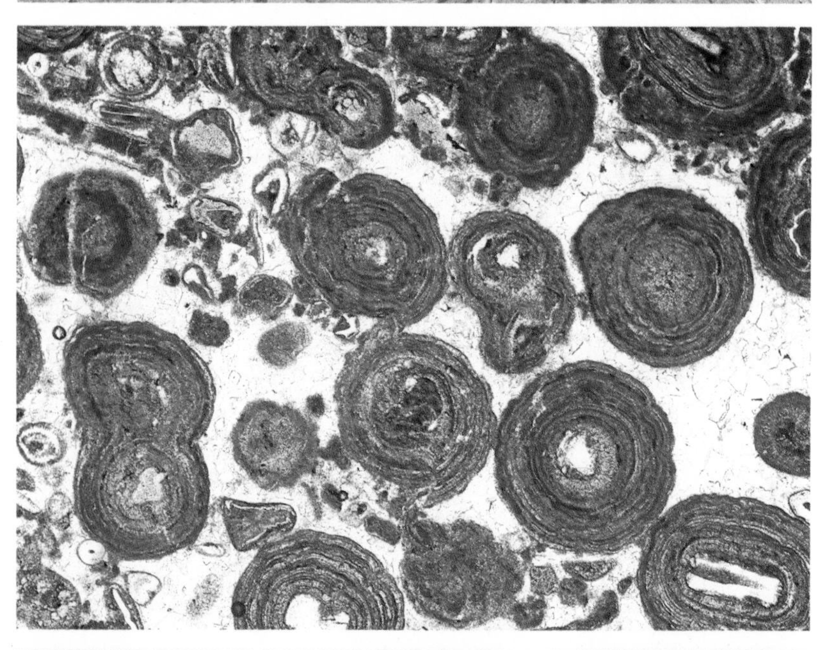

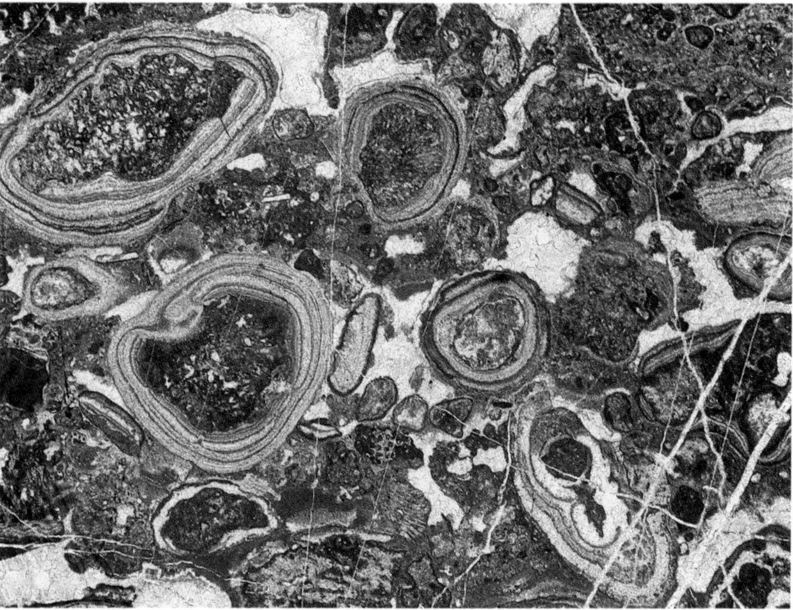

Pisoids and Oncoids

The nomenclature of carbonate grains which are larger than 2 mm in diameter and have an outer layer with concentric laminae, depends to a certain extent on an interpretation of their origin. Thus the term *pisoid* or *pisolith* usually refers to grains presumed to have formed inorganically, usually in a subaerial environment. On the other hand, *oncoids*, or *oncoliths*, are presumed to be biogenic, blue-green algae on the grain surfaces, trapping and binding fine sediment particles.

81 is a photograph of a polished rock surface showing oncoids. Note the size of the grains, the asymmetrical growth and the wavy nature of many of the laminae, all features characteristic of oncoids. The bluish-grey areas are sparry calcite and the orange-brown areas are stained with iron oxides.

82 and **83** show concentrically-laminated grains whose origins are more difficult to interpret. **82** shows grains which are about 2 mm in diameter. The outer surfaces are not as smooth as most ooids, although the concentric lamination is very regular. Grains in the upper right show irregular outer coats of micrite and some particles have apparently grown together to form compound grains (e.g. lower left). This latter feature is unlikely to occur in ooids, where precipitated carbonate laminae are formed while the grain is held in suspension. These grains are therefore interpreted as oncoids. The cement is sparite. This photograph is of a thin section made by Sorby in 1849 and illustrates the high quality of his sections.

83 shows grains with a regular, well-defined concentric layering, in grains up to 5 mm in diameter. This is typical of inorganic growth and these grains may be pisoids. Pisoids are commonly fractured or broken. Broken pieces can be seen towards the top right of the photograph.

81: *Polished surface, Llanelly Formation, Lower Carboniferous, Blaen Onneu, South Wales; magnification × 1.8.*
82: *Unstained thin section, Wenlock Limestone, Silurian, Malvern Hills, England; magnification × 13, PPL.*
83: *Stained thin section, Lower Jurassic, Greece; magnification × 11, PPL.*

Skeletal particles (Bioclasts)

Introduction

Skeletal particles, or bioclasts, are the remains, complete or fragmented, of the hard parts of carbonate-secreting organisms. There is such a variety in the mineralogy. structure and shape of skeletal material that several books could be written on this subject alone.

When trying to identify bioclasts, the following features should be considered:

1. The overall shape and size of the particle.
2. The internal wall structure of the particle. Many structures are more easily visible with polars crossed than in plane polarized light. It is important to distinguish those bioclasts which were originally calcite and have well-preserved wall structures from those which were originally aragonite and have had their wall structure modified or replaced during the alteration to calcite.

In this section we have attempted to show the diversity of skeletal structures present in ancient limestones, concentrating on examples from groups which are particularly common or occur over a wide stratigraphic range. For more detailed descriptions and illustrations of skeletal particles readers are referred to Majewske (1969), Horowitz and Potter (1971), Bathurst (1975) and Scholle (1978).

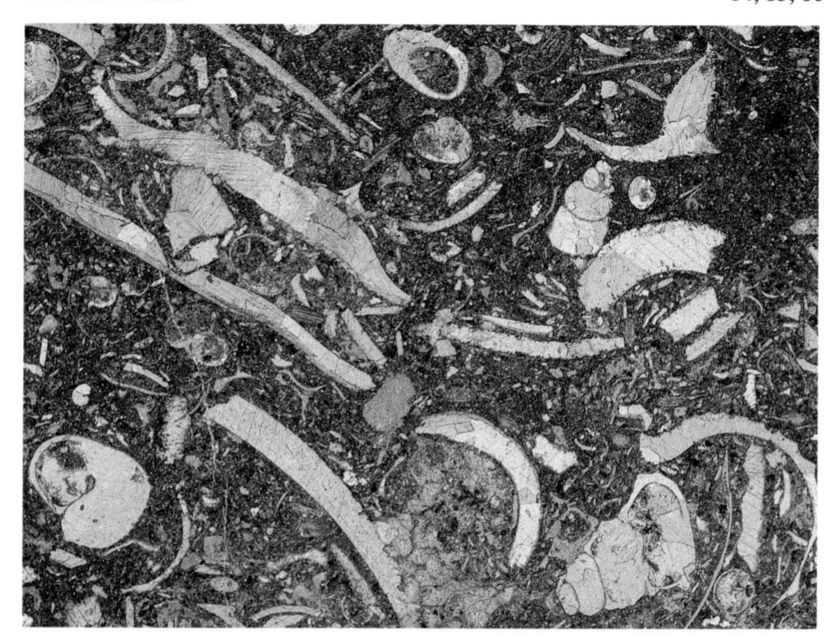

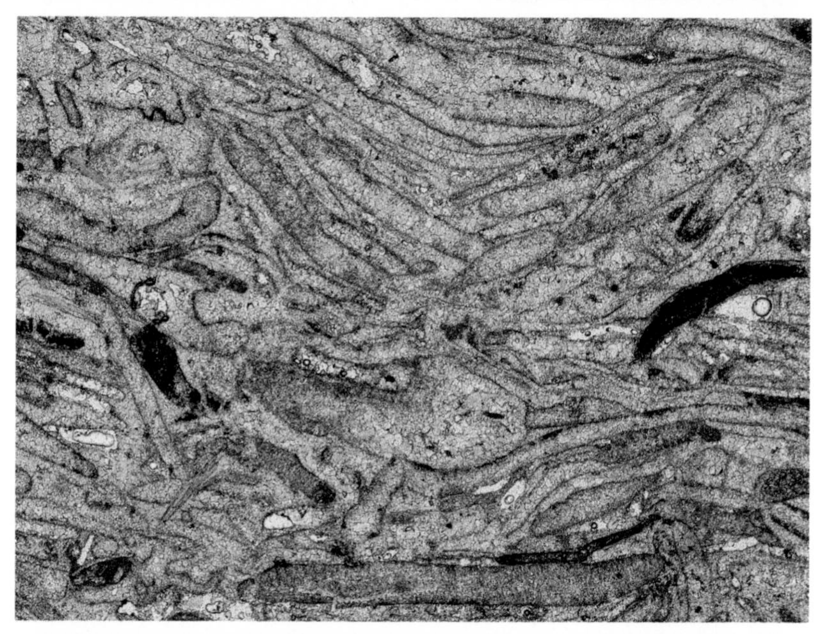

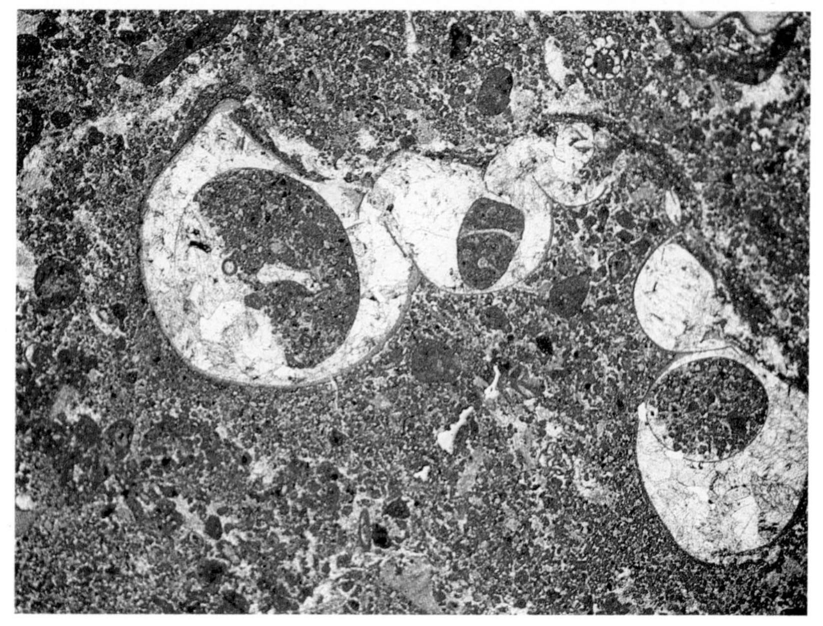

Bioclasts

Molluscs

Bivalves and gastropods are common components of limestones. Most were made of aragonite, so although there are a diversity of structures, these are not seen in ancient limestones. Most originally aragonitic molluscs are preserved as *casts* – that is the aragonite dissolved out during diagenesis leaving a mould which later became filled with a sparite cement. There are, however, important molluscan groups which had a calcite shell, especially among the bivalves, and these have well-preserved wall structures.

84 shows a limestone with abundant molluscan casts. In this case shell moulds have been infilled with a few large calcite crystals. Gastropods can be seen, both in long section (lower right) and transverse section (lower left). The long straight shells are bivalve fragments. Careful inspection shows that the long valves in the upper left have a two-layer structure – a thick layer of coarse sparite and a thin layer with a different structure. This latter layer may have been calcite originally, indicating that the organism had a mixed aragonite/calcite skeleton. The rock matrix is micritic sediment.

85 shows a limestone made up almost entirely of rounded bivalve fragments preserved as casts. The shape of the fragments is shown by the thin micrite rims on the margins of the shells. These are *micrite envelopes* and formed by micritization by endolithic algae (p. 54). The cement infilling the bivalves and between the shells is a fine sparite, initially pink-stained non-ferroan calcite, but becoming ferroan towards the centres of pore-spaces as indicated by the bluish staining.

86 illustrates a section through a large thick-shelled gastropod, again preserved as a cast. The outer margin of the shell is picked out by a thin calcite layer, not more than 0.5 mm thick at this magnification, but the inner margin is only clear where sediment has partially filled the internal cavity. The sediment around the shell contains abundant small peloids.

84: Stained thin section, Eyam Limestone, Lower Carboniferous, Ricklow Quarry, Derbyshire, England; magnification × 13, PPL.
85: Stained thin section, Upper Jurassic, Dorset, England; magnification × 14, PPL.
86: Stained acetate peel, Martin Limestone, Lower Carboniferous, Millom, Cumbria, England; magnification × 7, PPL.

Bioclasts

Molluscs *(continued)*

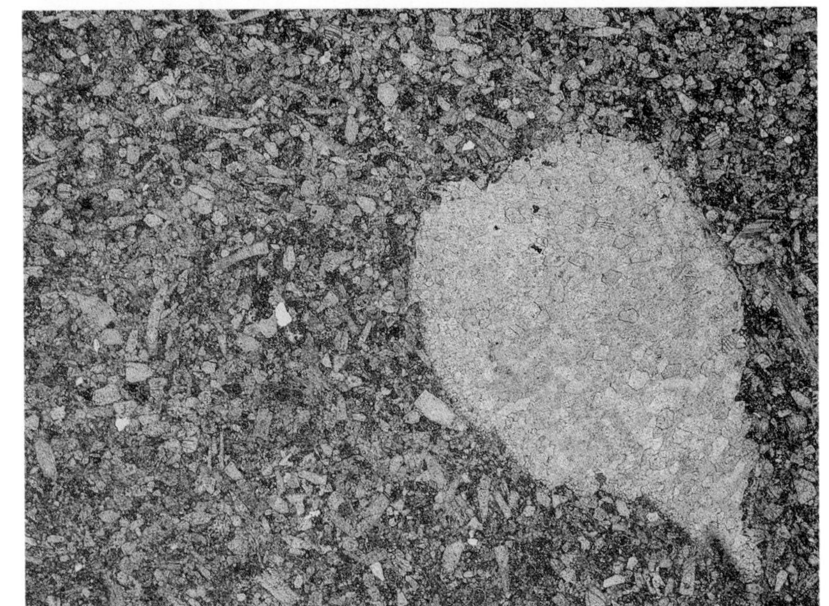

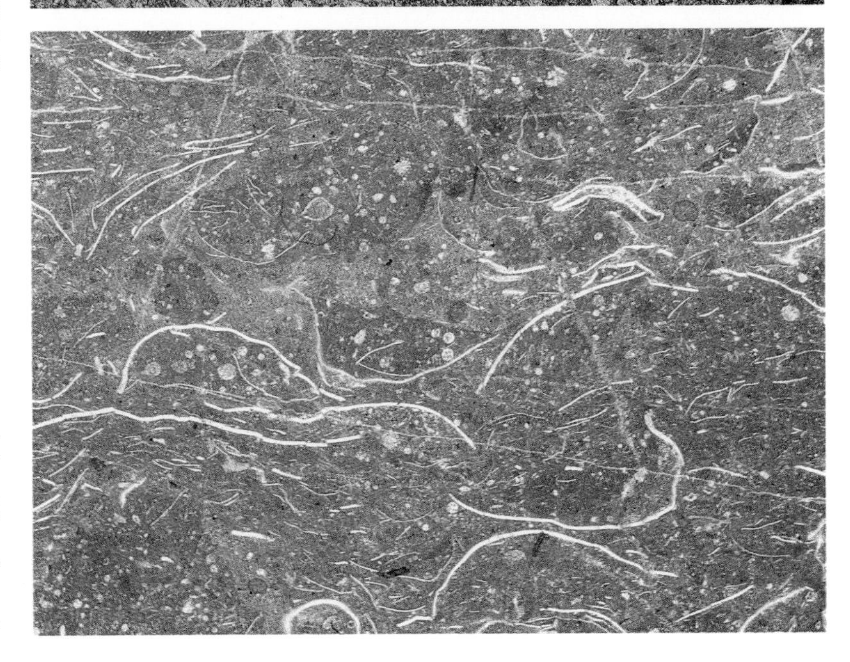

The photographs on this page illustrate bivalves which were entirely or partly calcite.

The oysters are one of the most important groups of calcitic bivalves. **87** shows two large pink-stained oyster fragments, each having a foliated internal structure. Fragments of oysters may be difficult to distinguish from brachiopods, although their thick shells with a rather irregular foliated structure are characteristic. Note also how the left-hand end of the upper fragment is upturned and splitting. The rest of the sediment comprises broken-up bioclasts set in a blue-stained ferroan calcite cement. The white areas are holes in the section.

Some bivalves have a thick prismatic layer, the prisms being elongated at right angles to the shell wall. **88** shows a fragment of the common Mesozoic bivalve *Inoceramus* (right). The shell is sectioned more or less parallel to its length and hence the prisms are seen in cross-section. Individual prismatic crystals break away easily from the shell and in this example most of the sediment is composed of these crystals, seen in various sections.

89 shows examples of thin bivalve shells known as filaments. These are the valves of planktonic bivalves and are common in Mesozoic pelagic limestones. The micritic sediment between the shells contains small circular areas of sparite. These are probably calcite casts of the siliceous microfossils, radiolaria (p. 82).

87: Stained thin section, Inferior Oolite, Middle Jurassic, Leckhampton Hill, Gloucestershire, England; magnification × 8, PPL.
88: Stained thin section, Upper Cretaceous, Strathaird, Skye, Scotland; magnification × 14, PPL.
89: Stained thin section, Triassic, Greece; magnification × 16, PPL.
Other molluscs are shown in **105**, **124**, **135**, **136**, **143**, **153**, **156** *and* **159**.

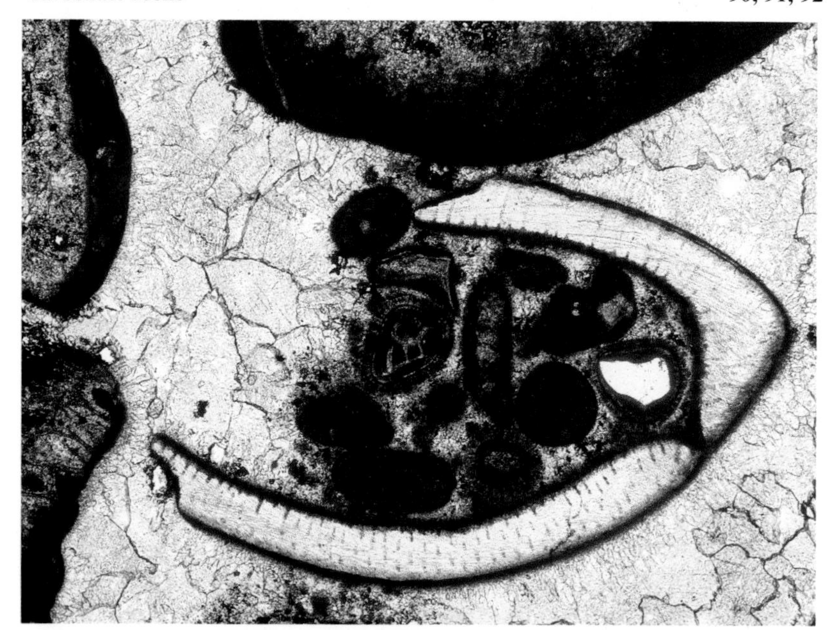

Bioclasts

Brachiopods

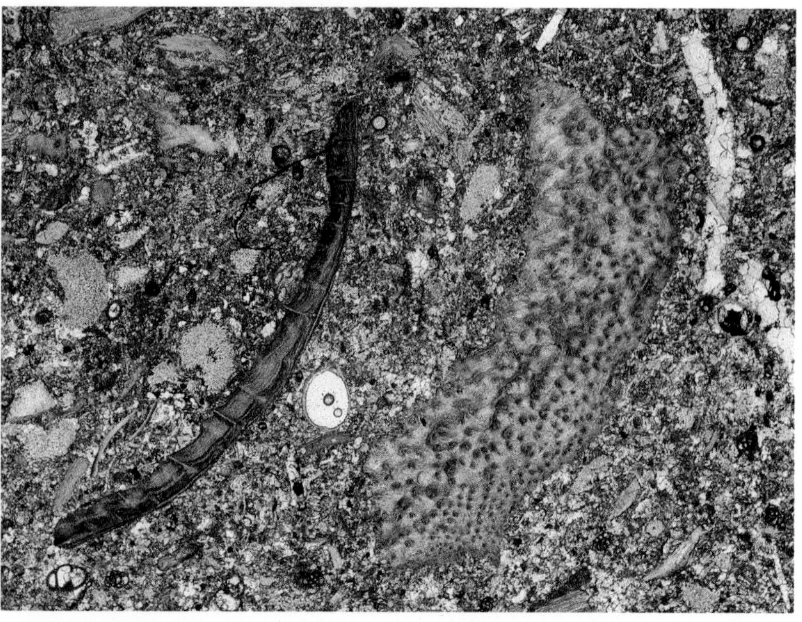

The articulate brachiopods are important constituents of Palaeozoic and Mesozoic limestones. They were originally calcite and so their shell structures are well-preserved. Typically, brachiopods have a thick inner layer of calcite fibres aligned with their length at a low angle to the shell wall. A thin outer prismatic layer may be preserved.

90 shows a broken brachiopod of which parts of both valves are present and surrounded by a micrite envelope (p. 54). The fibrous structure is clearly visible, as are fine tubes at right angles to the shell wall, filled with blue-stained ferroan calcite cement. These are *endopunctae* and they characterize some groups of brachiopods. The sample also shows a good example of coarse, blue-stained ferroan calcite cement.

91 shows two large fragments of *pseudopunctate* brachiopods. In these, the fibrous wall structure is interrupted, not by open tubes but by calcite rods. The left-hand fragment shows the pseudopunctae sectioned parallel to their length. Note the wavy nature of the fibres adjacent to the pseudopunctae. The right-hand fragment is a section of a shell showing the pseudopunctae in cross-section.

92 illustrates a brachiopod fragment with its outer prismatic layer preserved. The foliated nature of the inner part of the wall is also well shown. The shape of the fragment suggests that it is part of a ribbed shell. It is also *impunctate*, lacking either endopunctae or pseudopunctae. These factors in an Upper Jurassic brachiopod indicate that it is part of a rhynchonellid. The fine-grained calcite matrix contains abundant colourless fine sand- and silt-size quartz.

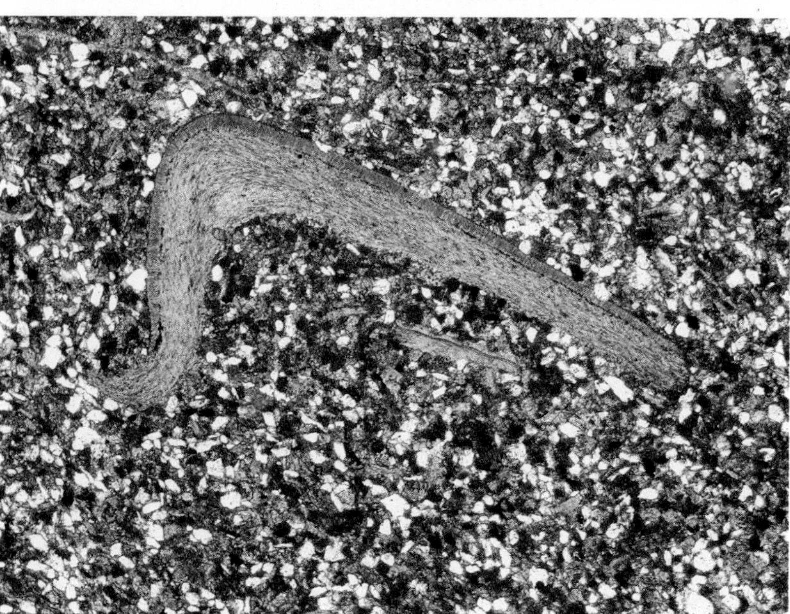

90: Stained thin section, Inferior Oolite, Middle Jurassic, Leckhampton Hill, Gloucestershire, England; magnification × 32, PPL.
91: Stained thin section, Monsal Dale Limestone, Lower Carboniferous, Cressbrook Dale, Derbyshire, England; magnification × 16, PPL.
92: Stained thin section, Upper Jurassic, Jebel Amsitten, Morocco; magnification × 40, PPL.

Bioclasts

Brachiopods *(continued)*

93 shows a number of small impunctate brachiopods with the large pedicle valve and smaller brachial valve complete. The roughly elliptical fragment in the lower centre is a section transverse to the length of the fibres making up the shell wall, and shows a characteristic fine net-like structure.

Some spiriferoid and pentameroid brachiopod shells have an inner layer composed of calcite prisms aligned at right angles to the length of the shell. In this case the outer foliated and prismatic layers are much reduced in thickness. **94** shows two large shell fragments with a thick inner prismatic layer. One is below the centre of the field of view, lying parallel to the bottom of the photograph, the other is near the left-hand edge. The sediment also contains brachiopod fragments with the more normal, foliated structure in a matrix of fine sparite, probably of neomorphic origin (see p. 60) containing grey-coloured crystals of replacement dolomite.

Some pseudopunctate brachiopods possess hollow spines. **95** shows transverse sections through several spines. They have a structure similar to the brachiopod valve with a foliated inner layer and an occasionally preserved outer prismatic layer. The section through the large spine in the upper left of the picture shows part of the outer prismatic layer preserved. Note how the shape of the spine gives the foliated layer a concentric structure. A longitudinal section through a brachiopod spine can be seen in the upper left-hand corner of **94**.

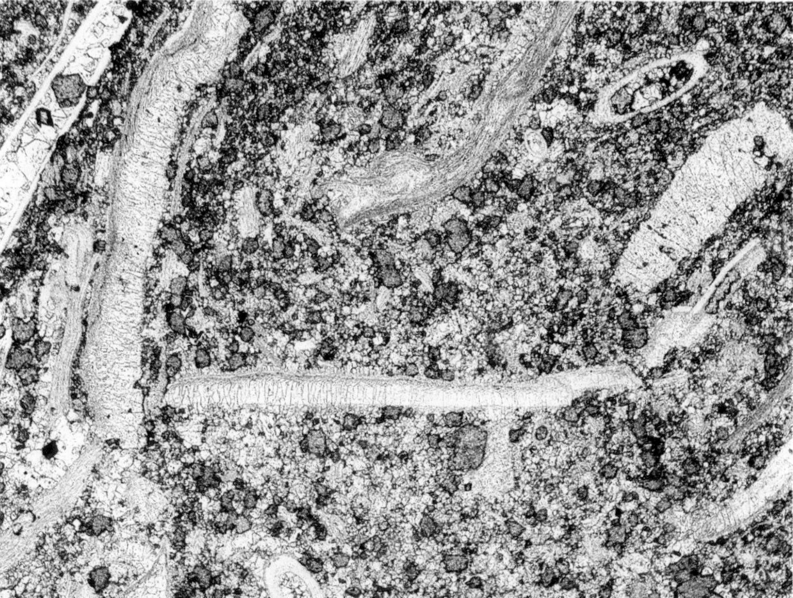

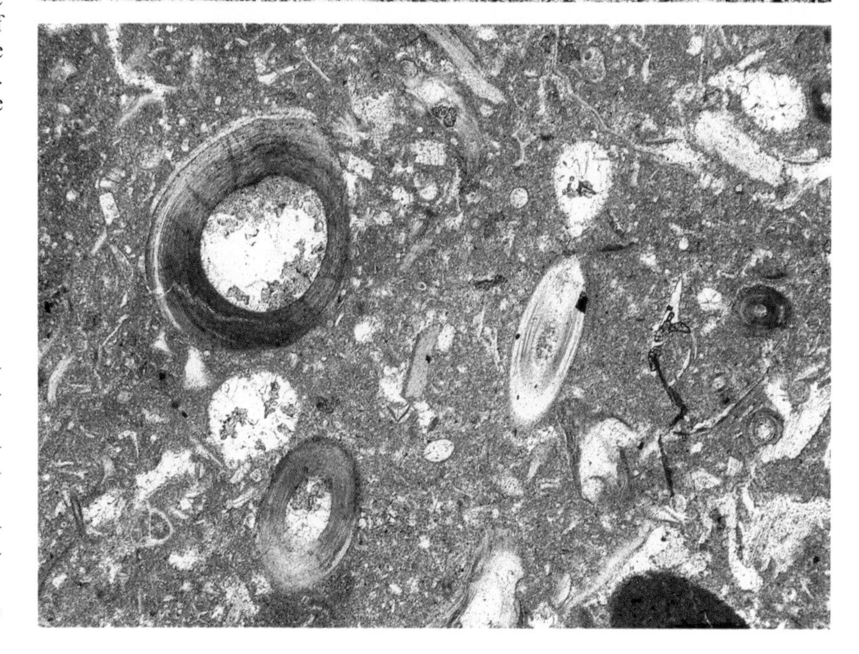

93: Stained acetate peel, Eyam Limestone, Lower Carboniferous, Ricklow Quarry, Derbyshire, England; magnification × 20, PPL.
94: Stained acetate peel, Eyam Limestone, Lower Carboniferous, Headstone Cutting, Derbyshire, England; magnification × 27, PPL.
95: Stained thin section, Eyam Limestone, Lower Carboniferous, Ricklow Quarry, Derbyshire, England; magnification × 28, PPL.
Other brachiopods and brachiopod spines are shown in **77**, **103**, **106**, **120**, **123** *and* **183**.

Bioclasts

Echinoderms

Echinoderms, particularly echinoids and crinoids, are major contributors to the allochemical fraction of marine limestones. They are easy to identify because they break down into plates which, although they may exhibit a wide variety of shapes, are single calcite crystals with uniform extinction. They usually have a speckled or dusty appearance as the result of infilling of the fine pores which permeate the plates.

96 and **97** show a crinoidal limestone in which the sediment is 75% crinoids. Note the speckled appearance of the plates, most of which have uniform interference colours and are thus single crystals, although the ossicle in the upper left comprises two crystals, one showing a greenish colour and one a red colour under crossed polars. The clear spar surrounding some of the crinoid fragments is a cement. The XPL photograph shows that the interference colour of this cement is the same as the adjacent crinoid fragment. Hence it is probable that the cement is in optical continuity with the crinoid. Such cements are common in echinoderm-bearing sediments and are called *syntaxial rim cements* (p. 57). The remainder of the sample comprises micritic sediment and fragments of fenestrate bryozoans (e.g. lower right hand corner).

Echinoid spines are widespread, particularly in Mesozoic and Cenozoic limestones. **98** shows one complete transverse section of a spine (lower right of field), together with a smaller broken fragment. Echinoid spines are circular or elliptical in cross-section and show a variety of radial structures. Like other echinoderm fragments, they are single crystals.

96 and 97: Stained thin section, Eyam Limestone, Lower Carboniferous, Once-a-week Quarry, Derbyshire, England; magnification × 13; ***96*** *PPL,* ***97*** *XPL.*
98: Stained thin section, Quaternary, Cap Rhir, Morocco; magnification × 31, PPL.
Echinoderm fragments are also shown in **73**, **76**, **78**, **132**, **133**, **139**, **148**, **154**, **178**, **183** *and* **184**.

Bioclasts

Corals

Corals are best identified by their overall morphology. The rugose and tabulate Palaeozoic corals were calcite, thus their microstructures are well-preserved. The walls are usually fibrous and small fragments which lack evidence of the characteristic coral form can be difficult to identify.

99 shows a transverse section and parts of two longitudinal sections of the colonial rugose coral *Lithostrotion*. Note the thick outer wall and septa seen in the transverse section. The columella and thin tabulae are clearly visible in the longitudinal section. Parts of the coral walls have been silicified (brownish colour). The pore-filling material is mainly sparite cement with some micritic sediment between the corallites.

100 shows a section through a tabulate coral. Note the corallite walls and thin tabulae but absence of other internal structures. The infill is sparite cement, initially non-ferroan calcite (pink-stained), but finally ferroan (blue-stained).

The Mesozoic and Cenozoic scleractinian corals are composed of aragonite and hence their microstructure is not well-preserved in limestones. Scleractinian corals are shown in **126**, **144** and **145**.

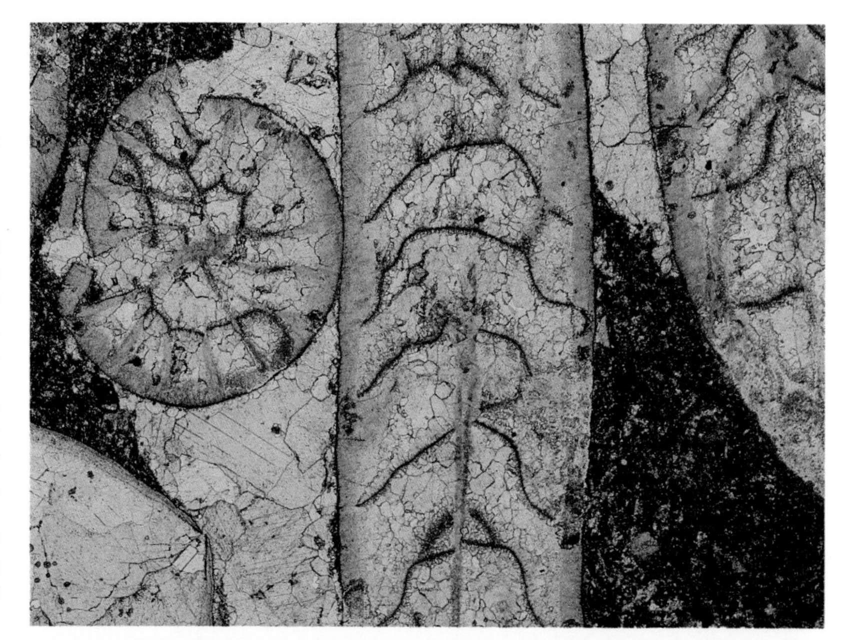

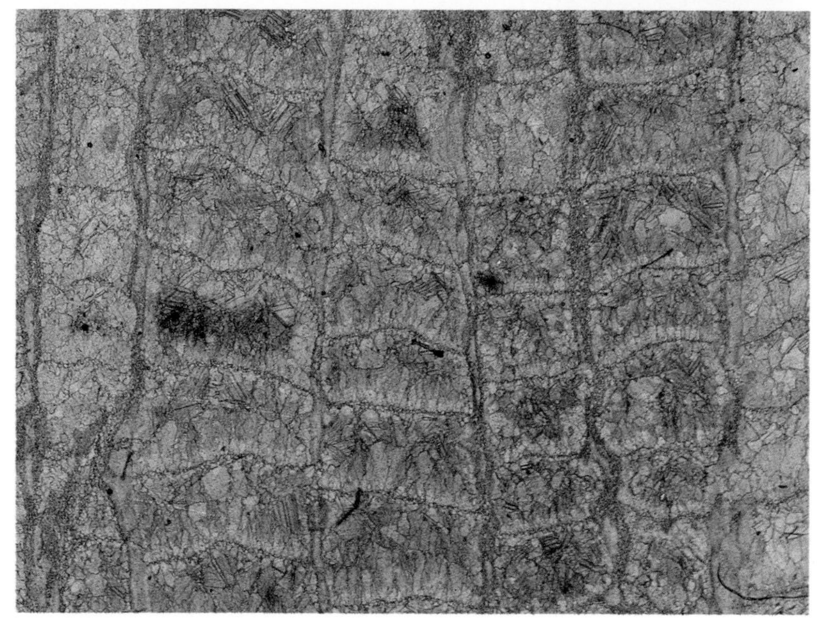

***99**: Stained thin section, Monsal Dale Limestone, Lower Carboniferous, Coombs Dale, Derbyshire, England; magnification × 16, PPL.*

***100**: Stained thin section, Torquay Limestone, Devonian, Brixham, Devon, England; magnification × 16, PPL.*

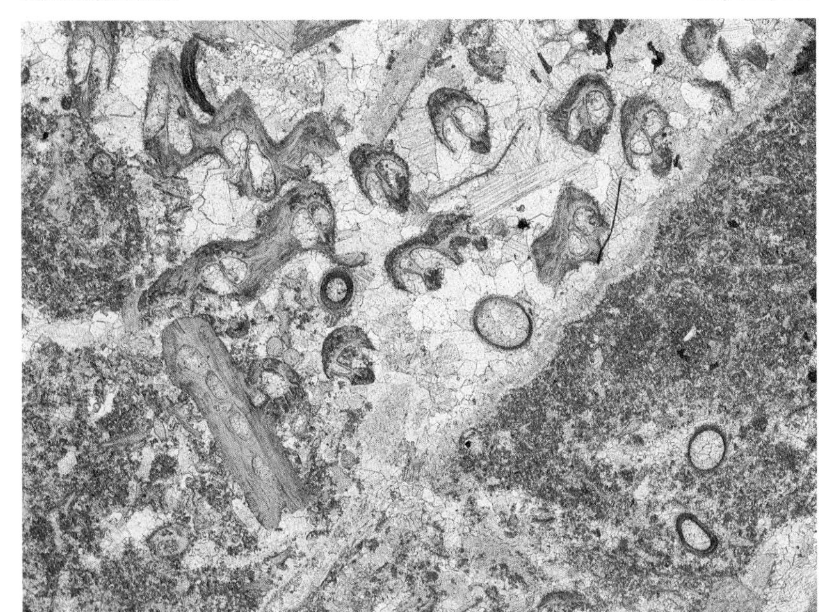

Bioclasts

Bryozoans

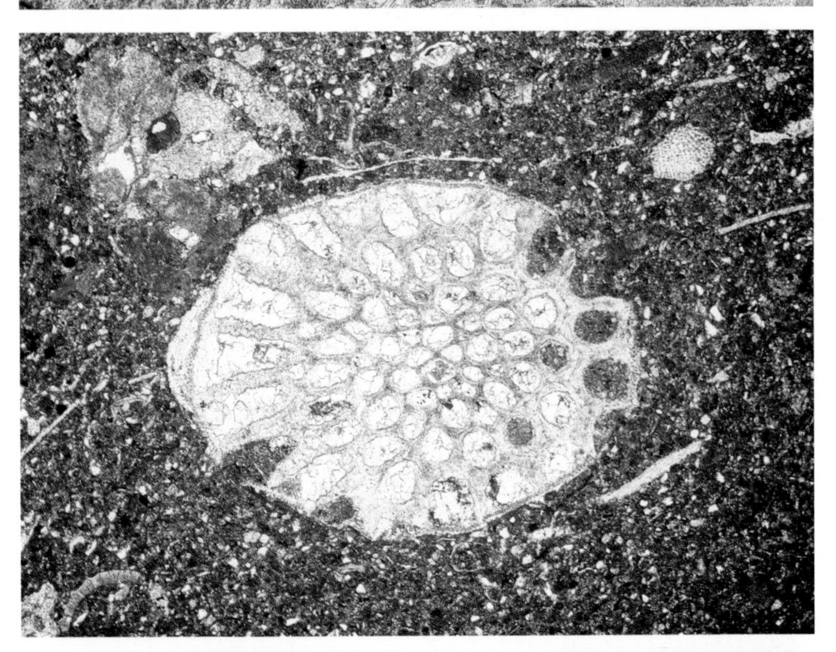

Bryozoans are widespread in marine limestones and are particularly common in Palaeozoic reef complexes. Most bryozoans had calcite hard parts and a laminated wall structure is preserved.

Among the most characteristic bryozoans are the frond-like fenestrate types, examples of which are seen in **101**. Note the thick wall of laminated calcite surrounding cement-filled pores (zooecia). Most of the fragments are transverse sections but the large piece to the lower left of centre is a longitudinal section.

102 is a transverse section of a stick-like bryozoan colony, showing the overall rounded shape of the 'stem' and of the zooecia within. Some of these have been infilled with fine sediment (upper right of fragment) but most have a blue-stained, ferroan calcite cement infill.

In **103**, the two circular, concentrically-laminated grains stained red-brown are brachiopod spines. These are encrusted by a bryozoan. Note the thick calcite wall of the bryozoan and the pores of different sizes within the skeleton, filled with pink-stained non-ferroan calcite cement. Some fragments of fenestrate bryozoans can be seen along the left-hand side of the photograph.

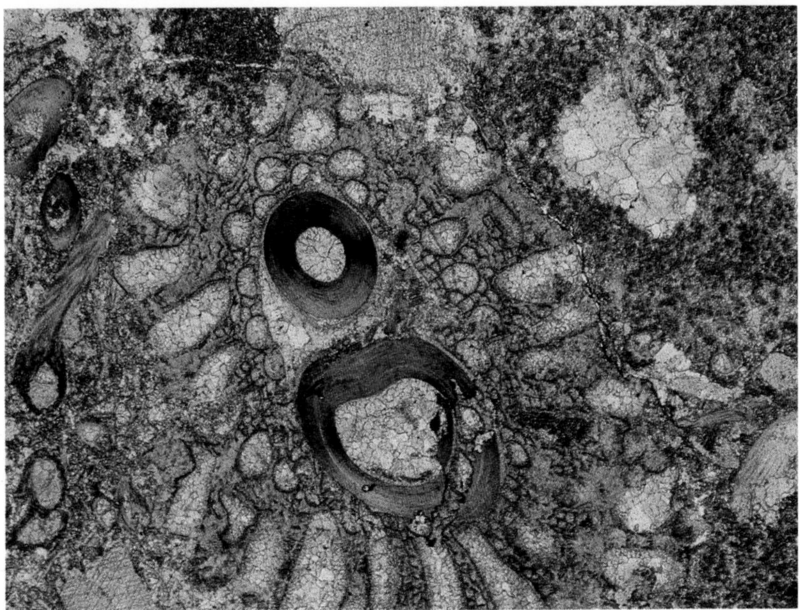

101: Stained thin section, Eyam Limestone, Lower Carboniferous, Ricklow Quarry, Derbyshire, England; magnification × 16, PPL.
102: Stained thin section, Ouanamane Formation, Middle Jurassic, Western High Atlas, Morocco; magnification × 27, PPL.
103: Stained thin section, Red Hill Oolite, Elliscales Quarry, Dalton-in-Furness, Cumbria, England; magnification × 20, PPL.
Other bryozoans are shown in **96**, **97**, **132**, **133** *and* **178**.

Bioclasts

Arthropods

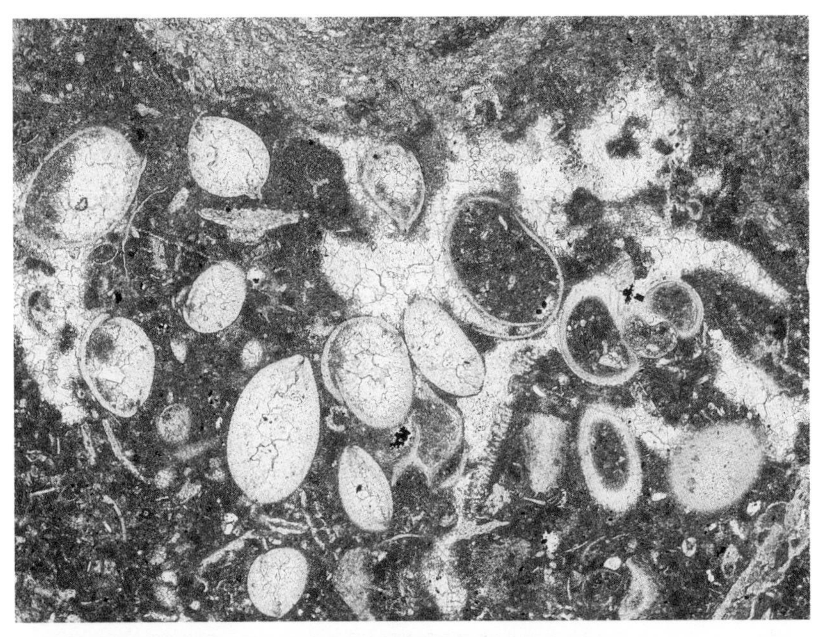

Ostracods

These photographs show examples of the arthropod microfossils, the ostracods, which are widespread particularly in sediments deposited in brackish or hypersaline conditions. Ostracods have thin valves with a finely prismatic or granular microstructure.

104 shows a group of complete two-valved shells, some filled with sparite cement, some with micritic sediment and some with both. Note the overlap of valves seen in some sections – a characteristic feature of many ostracods.

105 shows disarticulated ostracod valves (thin curved shells) associated with longer straight lengths of shell, which are fragments of a calcitic non-marine bivalve.

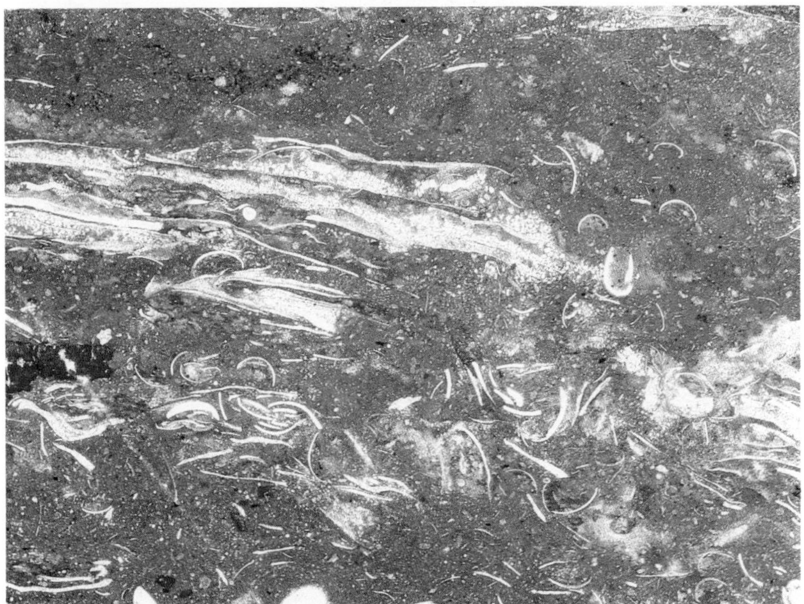

104: Stained thin section, Red Hill Oolite, Lower Carboniferous, Elliscales Quarry, Dalton-in-Furness, Cumbria, England; magnification × *40, PPL.*
105: Unstained thin section, Upper Carboniferous, Cobridge Brickworks, Hanley, Staffordshire, England; magnification × *16, PPL.*
Ostracods are shown also in **117**, **119** *and* **136**.

Trilobites

Trilobite hard parts were originally calcite and a finely granular microstructure is preserved. Each crystal is in a similar but not identical orientation to its neighbours, leading to sweeping extinction when the sample is rotated with the polars crossed (not illustrated here).

106 shows a cross-section of a trilobite (centre) and part of a brachiopod shell (base). Note the hooked shape seen at the left-hand end of the trilobite fragment, produced by incurving of the skeleton at its margin. A vein of blue-stained ferroan calcite follows the edge of the skeleton along part of its length. Note that the trilobite is stained mauve and hence consists of slightly ferroan calcite. This contrasts with the brachiopod fragment which is non-ferroan calcite. In some rocks it is thought that bioclasts originally comprising high magnesium calcite may be replaced by ferroan calcite, whereas those of low magnesium calcite remain unaffected.

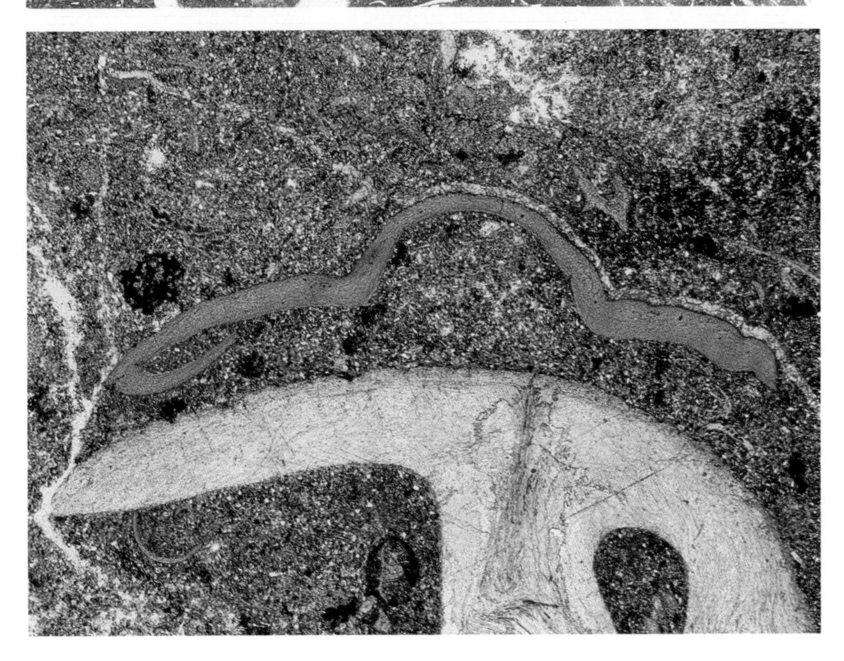

106: Stained thin section, Wenlock Limestone, Silurian, England; magnification × *21, PPL.*

Bioclasts

Foraminifera

Foraminifera are widespread in marine limestones. Most are calcite but they show a variety of shapes and wall structures. A selection of examples showing some of the variation amongst the foraminifera, is shown here.

The largest and perhaps the best-known foraminifera are the nummulites of the Lower Tertiary, examples of which are shown in **107**. Note the thick walls which have a radial fibrous structure, the fibres being aligned at right angles to the test wall. The matrix is mainly micritic sediment with a little blue-stained ferroan calcite cement.

108 shows discocyclinids, a type of foraminifer with many shall chambers. The matrix is micrite with many fragmented bioclasts.

109 shows a foraminiferal limestone in which the organisms are micrite-walled miliolids. The cement is fine sparite although unfilled pore-spaces remain (e.g. centre of field of view). Partly-filled moulds of bivalves can be seen outlined by thin micrite envelopes. These are the elongate curved grains seen on the right-hand side of the photograph.

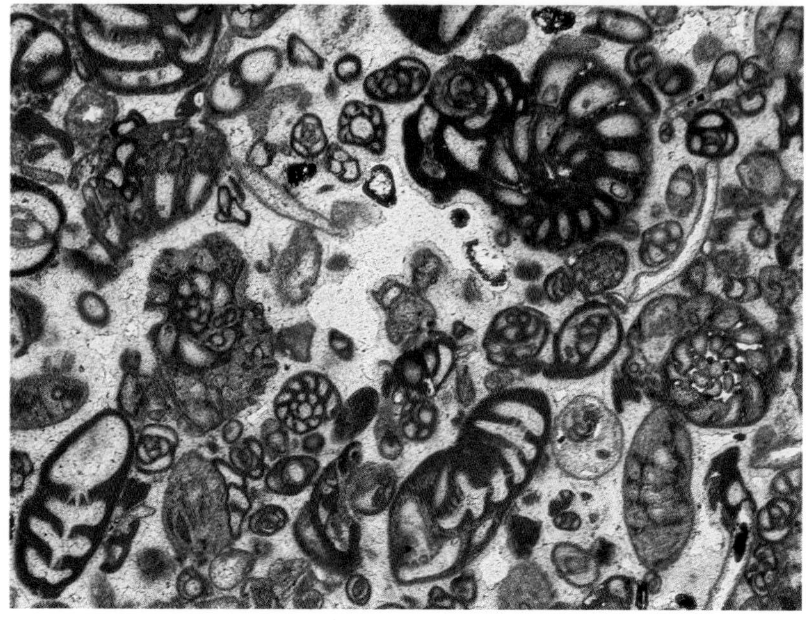

107: Stained thin section, Eocene, San Salvador, Majorca; magnification × 15, PPL.
108: Stained thin section, Eocene, Greece; magnification × 16, PPL.
109: Stained thin section, Upper Miocene, Cala Pi, Majorca; magnification × 27, PPL.

Bioclasts

Foraminifera *(continued)*

110 shows a number of species of micrite-walled forams (endothyracids). Note how the different sections show different arrangements of chambers. Much of the sediment comprises fragmented bioclasts in a pink-stained, non-ferroan calcite matrix.

Foraminifera are important contributors to pelagic sediments. **111** shows numerous pelagic foraminifera in an almost opaque micritic sediment. The large keeled forms are globorotalids, and smaller types include rounded globeriginids.

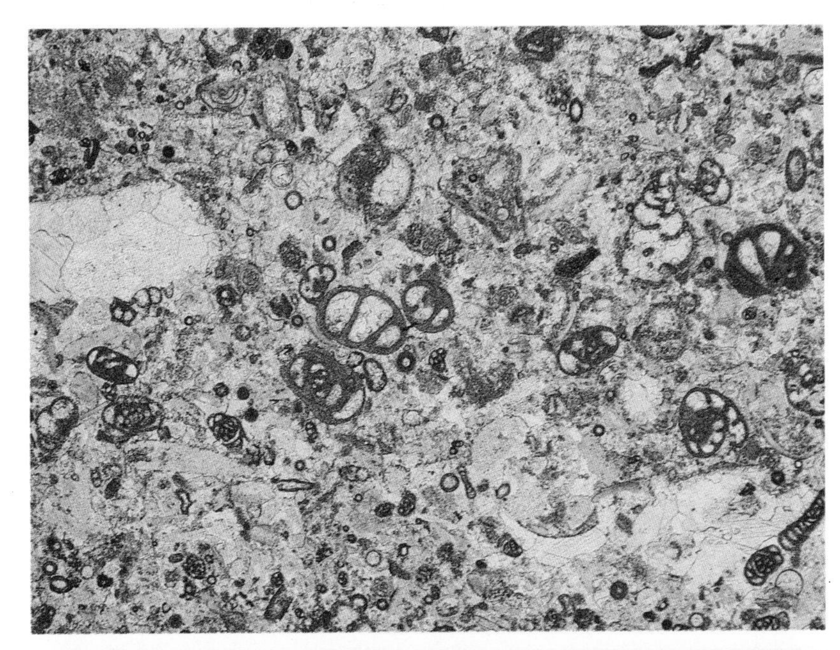

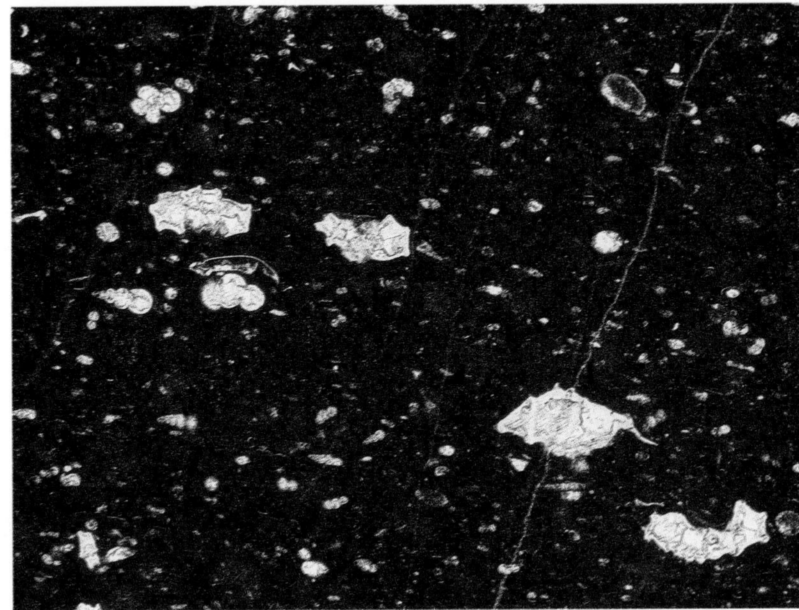

110: Stained thin section, Woo Dale Limestone, Lower Carboniferous, Dam Dale, Derbyshire, England; magnification × 19, PPL.
111: Stained thin section, Upper Cretaceous, Pindos Zone, Central Greece; magnification × 35, PPL.
Other foraminifera are shown in **116**, **120** *and* **157**.

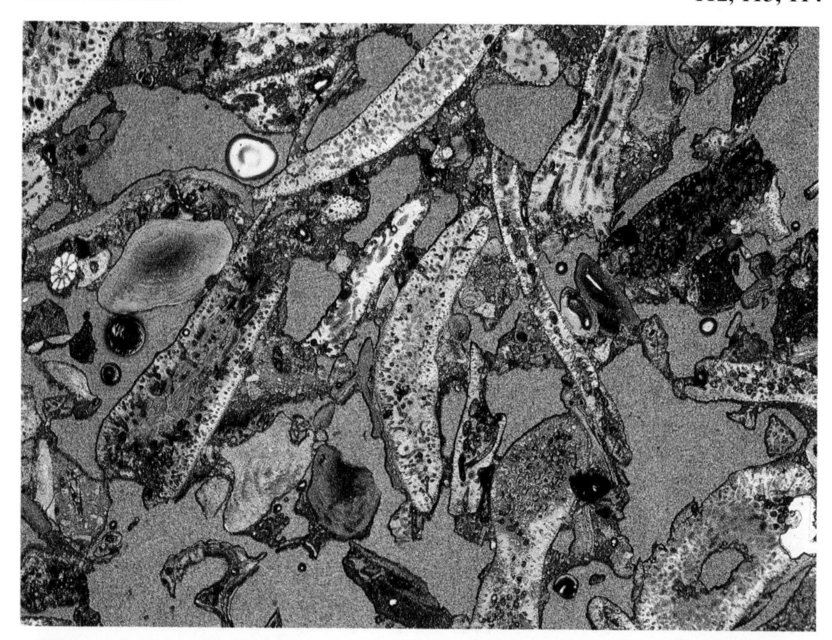

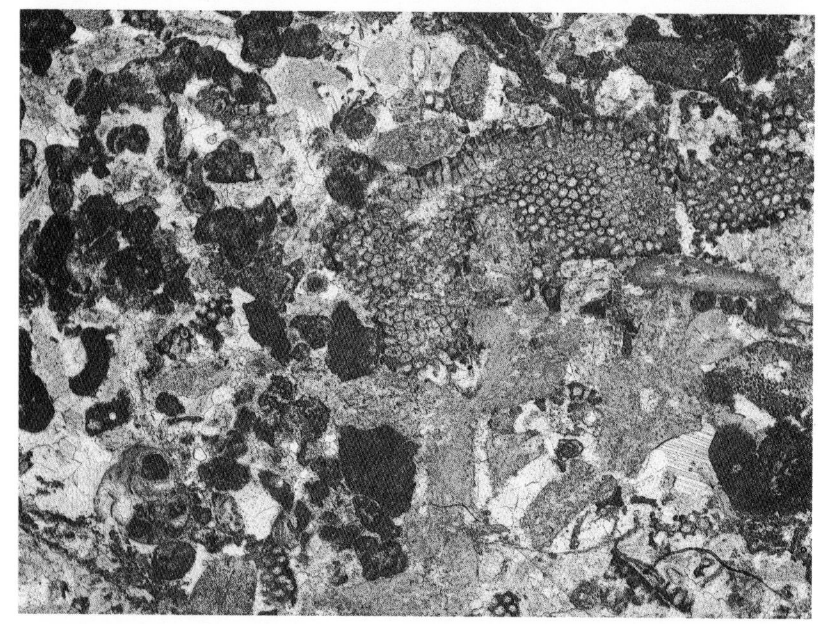

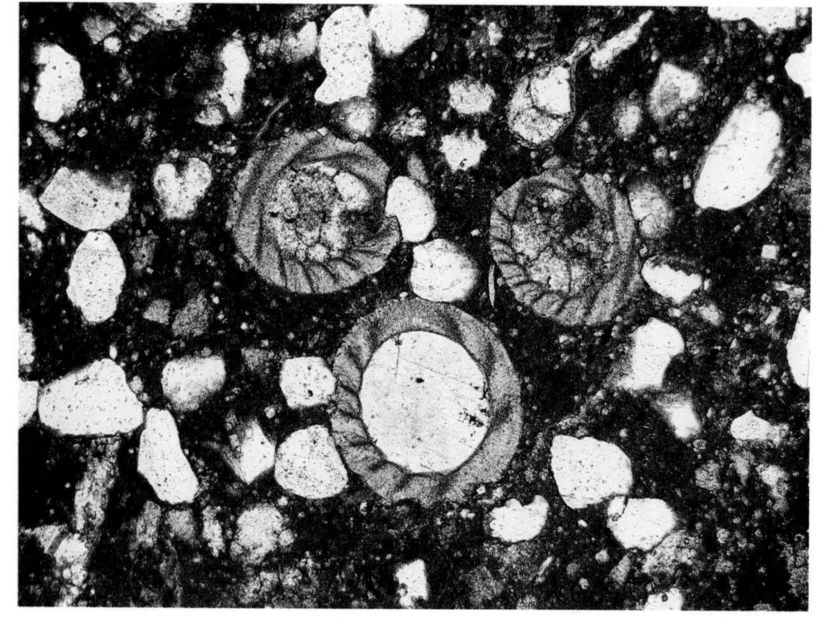

Bioclasts

Algae

Those algae in which all or part of the skeleton becomes calcified are known as the skeletal calcareous algae. They are important contributors to carbonate sediments throughout the Phanerozoic and exhibit a wide diversity of forms. Green algae are one of the most important groups and the photographs on this page illustrate three examples, one from each of the major groups, the Codiaceae, Dasycladaceae and Charophyceae. For details of calcareous algae see Johnson (1961), Wray (1977) and Flügel (1982).

112 shows segments of one of the common forms of codiacean alga, *Halimeda*, which still occurs today. Living examples contain organic filaments embedded in aragonite. The example shown is from a poorly-consolidated Quaternary sediment which had to be impregnated with resin before a peel could be made. The grey areas between the algal segments and in the holes originally occupied by the filaments are the impregnating medium. In this sample the *Halimeda* segments are still aragonite, although the wall structure cannot be seen at this magnification. *Halimeda* fragments are often poorly-preserved because of the loss of microstructure during the replacement of aragonite by calcite.

113 shows two types of algae. The large fragment with the honeycomb structure and walls of fine-grained calcite is the common Carboniferous dasycladacean alga *Koninckopora*. Several algal segments of a different type can be seen below the *Koninckopora*. At the level of viewing here they have no discernible wall structure and there is a slight resemblance to echinoderm fragments. In fact they have a fine fibrous wall structure and hence are not single crystals. They may show branching and an example of Y-branching can be seen in the lower-right part of the photograph. These belong to a problematic group, often referred to as ancestral coralline algae, but sometimes classified with the foraminifera or stromatoporoids.

The third group of green algae are the charophytes, although these are sometimes classified separately. They are freshwater plants, occurring in the Mesozoic and Cenozoic, and usually only the reproductive parts (oogonia) are calcified. These are small egg-shaped bodies with various ornaments. **114** shows three oogonia in cross-section.

112: Stained acetate peel, Quaternary, Mombasa, Kenya; magnification × 13, PPL.
113: Stained thin section, Chee Tor Rock, Lower Carboniferous, Tunstead Quarry, Derbyshire, England; magnification × 17, PPL.
114: Stained thin section, Iggui el Behar Formation, Upper Jurassic, Western High Atlas, Morocco; magnification × 56, PPL.

Bioclasts

Algae *(continued)*

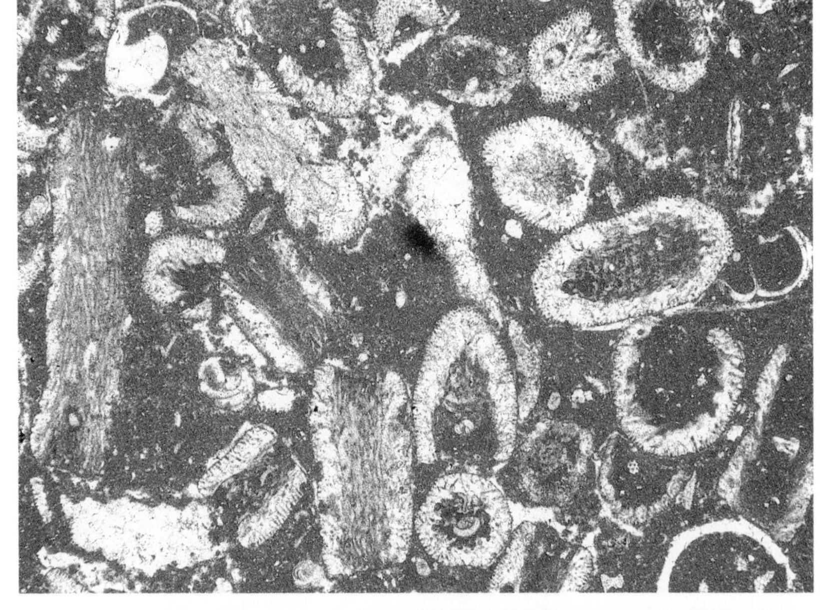

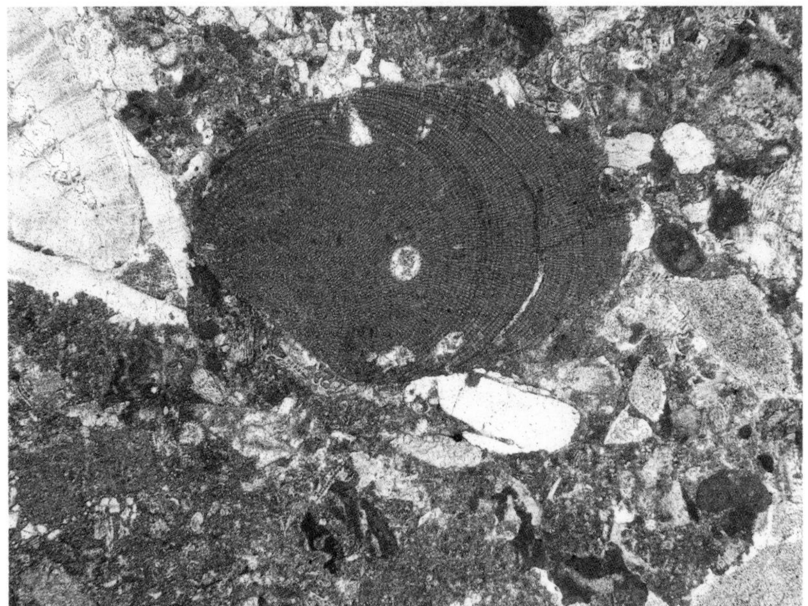

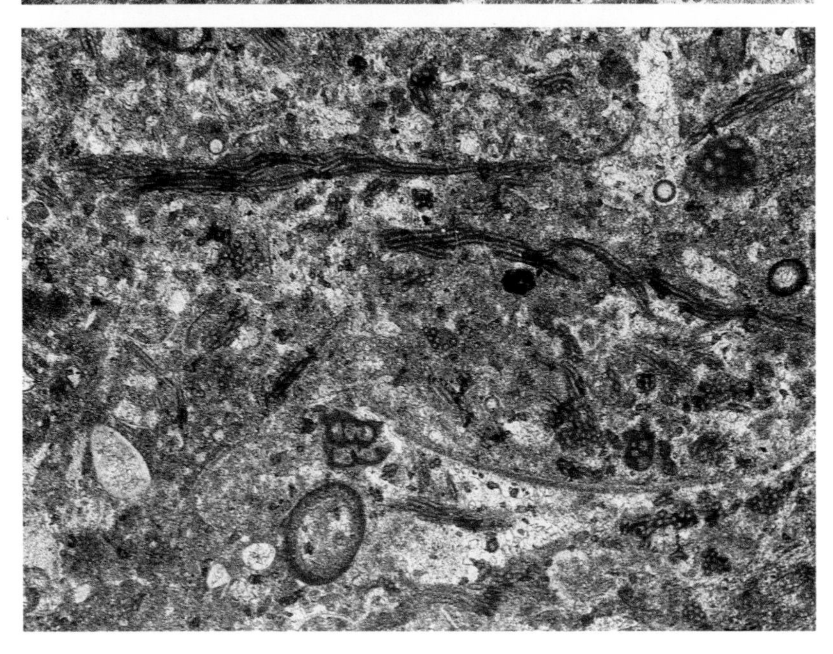

Many algae possess a central stem encased with calcium carbonate, through which filaments pass to the outside. **115** shows numerous sections of such an alga in a brown-stained micrite matrix. Both longitudinal and transverse sections are present. The transverse sections are roughly circular or elliptical and the centres are infilled with micrite sediment. Around the margins of the wall can be seen the holes formerly occupied by the filaments, now occupied by micrite sediment. Details of the wall structure have not been preserved so the alga was probably aragonite. The longitudinal sections show that the centre of the stem contains poorly-preserved casts of the algal filaments.

The red algae are important skeletal calcareous algae, and one group, the coralline algae, are major contributors to sediments, including reefs, during the Cenozoic. **116** shows a fragment of a coralline alga, with its characteristic reticulate appearance caused by thin micrite walls separating small, more or less rectangular, cells. The spar-filled holes within the skeleton, called conceptacles, are also characteristic. To the left of the coralline alga can be seen part of a nummulitid foraminifera, with its thick radial-fibrous wall.

The blue-green algae occur typically as long narrow filaments and only a few species become calcified. *Girvanella*, illustrated in **117** is widespread and occurs over a long stratigraphic range. It is made up of bundles of narrow tubes about a millimetre in diameter at this magnification, with a thin micrite wall. They can be seen in longitudinal section (e.g. upper part of the photograph) and transverse section (e.g. lower right). The remainder of the sediment comprises a few bioclasts (e.g. an ostracod, lower left) and a mixture of carbonate mud sediment and sparite cement, the latter being partly pink-stained, non-ferroan calcite and partly bluish ferroan calcite.

115: Stained thin section, Upper Cretaceous, Tunisia; magnification × 19, PPL.

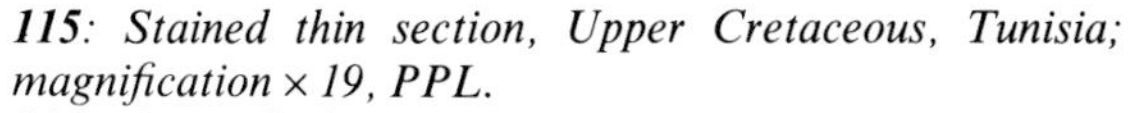

116: Stained thin section, Eocene, Greece; magnification × 23, PPL.

117: Stained thin section, Chatburn Limestone, Lower Carboniferous, Chatburn, Lancashire, England; magnification × 37, PPL.

Other algae are shown in **76**, **128**, **130** *and* **150**.

Bioclasts

Calcispheres and Worm Tubes

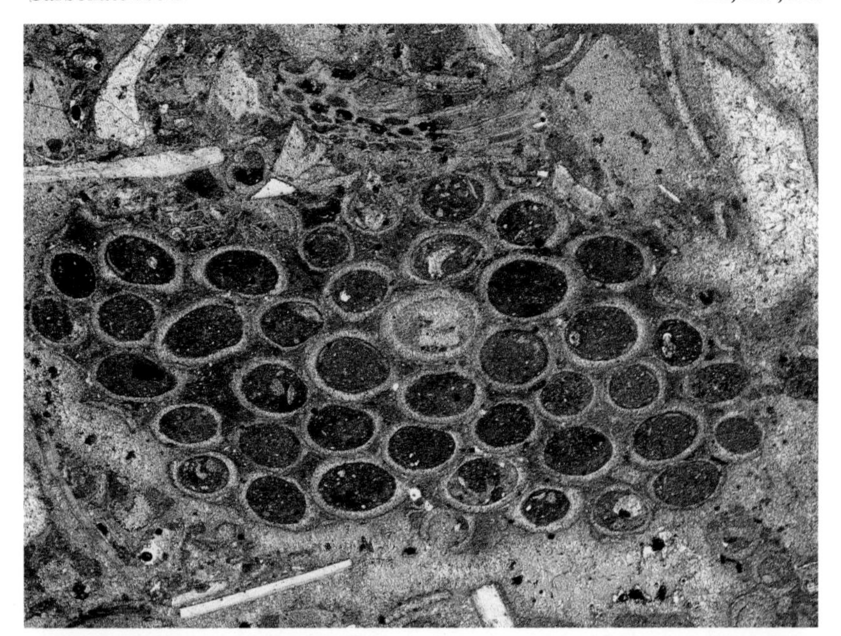

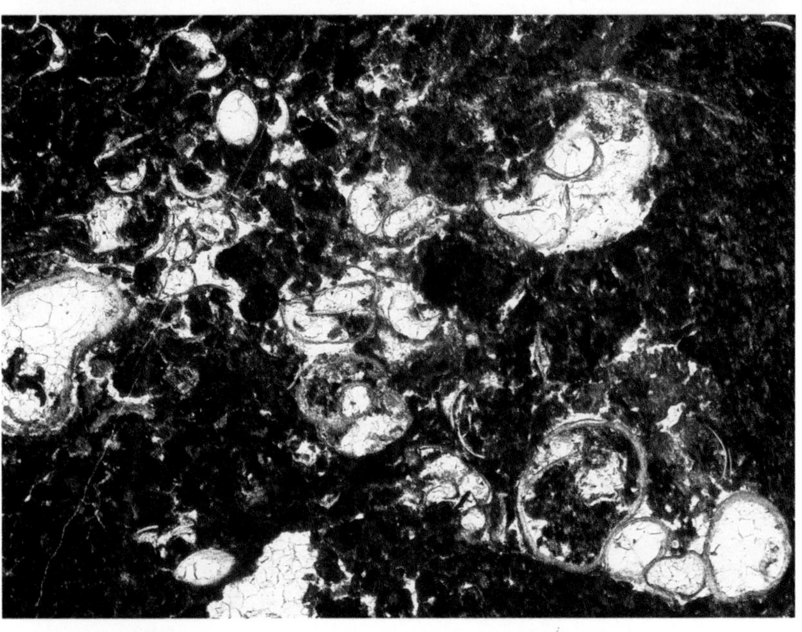

Worm tubes

Although rarely abundant, calcareous worm tubes are widespread in shallow marine and freshwater limestones. **118** shows a bioclastic limestone with a large fragment comprising numerous worm tubes, seen in cross-section. Most of the tubes are filled with micrite. The associated fauna include a bryozoan (top centre), molluscan casts (top right), an echinoderm fragment (top right) and brachiopods (bottom). The sediment is cemented by blue-stained ferroan calcite. **119** shows sections through the coiled calcareous worm tube *Spirorbis*. Sparry calcite cement fills the chambers and the surrounding sediment is micrite with a few thin-valved ostracods (e.g. upper left).

118: Stained thin section, Inferior Oolite, Middle Jurassic, Leckhampton Hill, Gloucestershire, England; magnification × 13, PPL.
119: Unstained thin section, Ardwick Limestone, Upper Carboniferous, Manchester, England; magnification × 17, PPL.
*Worm tubes are also shown in **214** and **215**.*

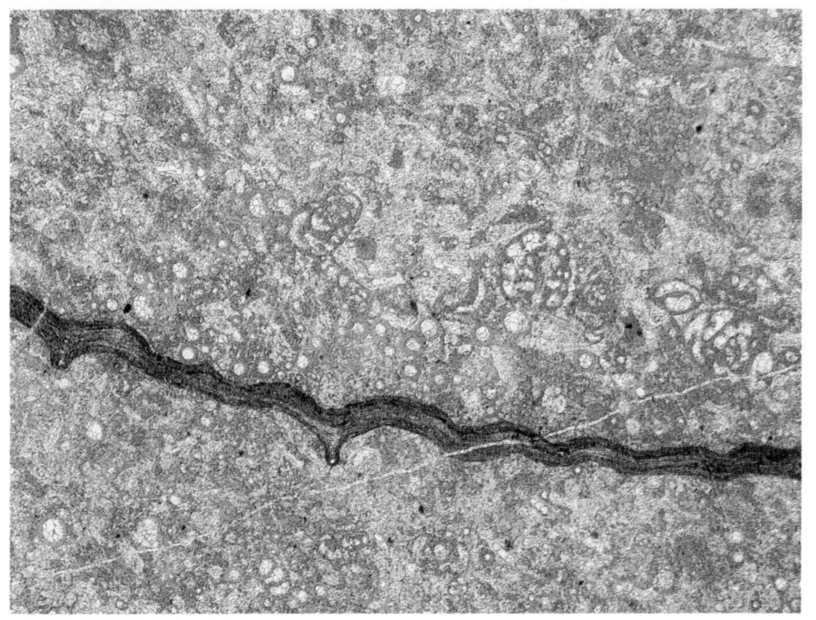

Calcispheres

Calcispheres are small hollow spherical bodies of calcite, usually with a micritic wall. They are particularly common in Upper Palaeozoic limestones and may be the calcified reproductive parts of dasycladacean algae. **120** shows numerous calcispheres – the circular objects with deep, red-brown-stained walls – associated with micrite-walled endothyracid foraminifera and a foliated brachiopod shell which extends right across the field of view.

120: Stained acetate peel, Woo Dale Limestone, Lower Carboniferous, Long Dale, Derbyshire, England; magnification × 21, PPL.

Non-skeletal algae

Stromatolites

Stromatolites are laminated rocks interpreted as fossilized algal mats. The mats are formed of filamentous blue-green algae. The laminae in stromatolites are usually alternations of carbonate mud and grainy, often pelleted, sediment. The laminae are, at most, a few millimetres thick and are often more easily seen in hand specimens than in thin sections. Laminoid fenestrae (p. 68) are often associated with stromatolites.

121 is a photograph of a polished block showing a stromatolite. Note that the layering is irregular and partly picked out by colour differences. The irregularity of the layering helps to differentiate laminated sediments formed from algal mats from those formed by physical processes. The laminae in stromatolites may form flat or crinkly structures or may build up into columns or domes. Concentric algal laminations about a nucleus give rise to the grains known as oncoids (p. 38).

122 shows a thin section of the same specimen as that illustrated in **121**. The laminations consist of alternating thin micritic layers and layers containing a mixture of micrite and sparite. In some areas the micrite has a vaguely pelleted structure which is characteristic of stromatolites. The more irregular micrite areas may have been coating the algal filaments which then decayed, leaving a mould which was later filled with sparite cement.

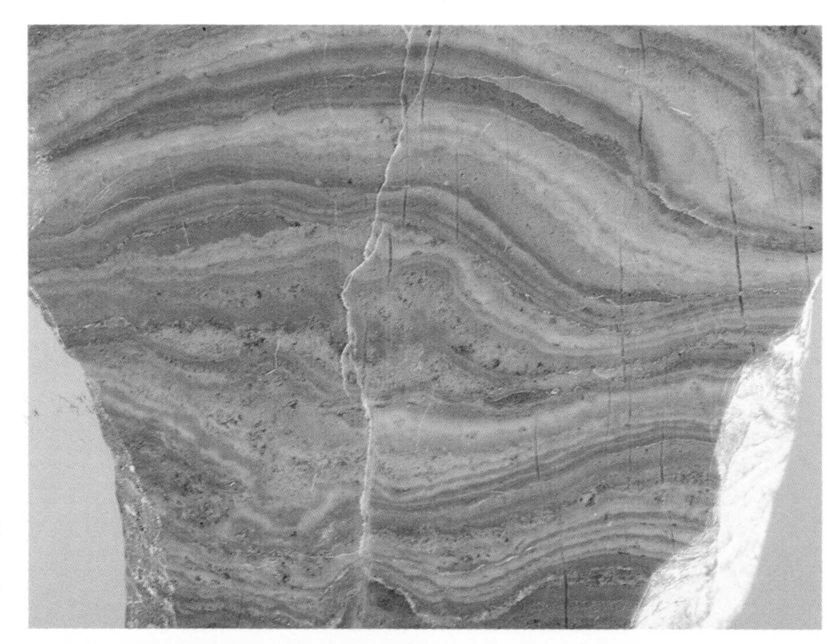

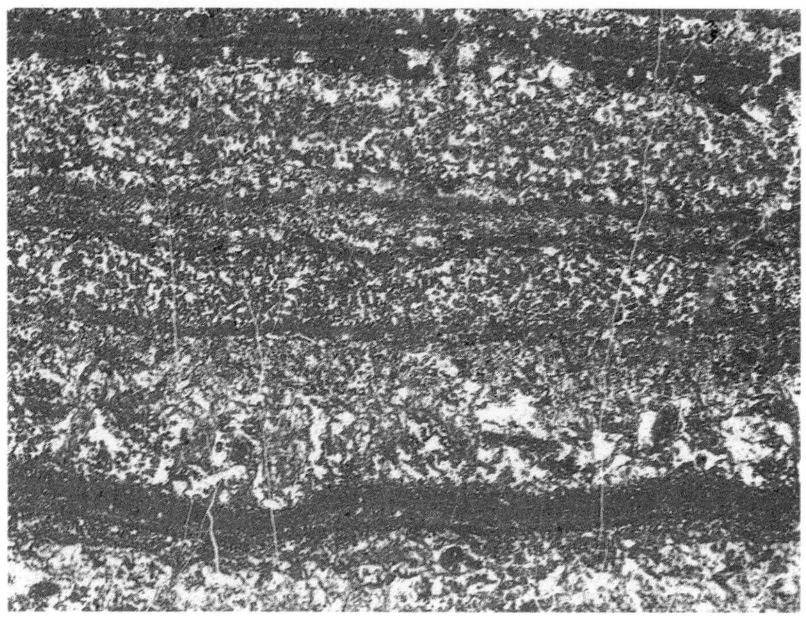

121 and 122: Lower Carboniferous, Carrière de la Vallée Heureuse, Boulonnais, France; 121 hand specimen, magnification × 1.8; 122 stained thin section, magnification × 12, PPL.

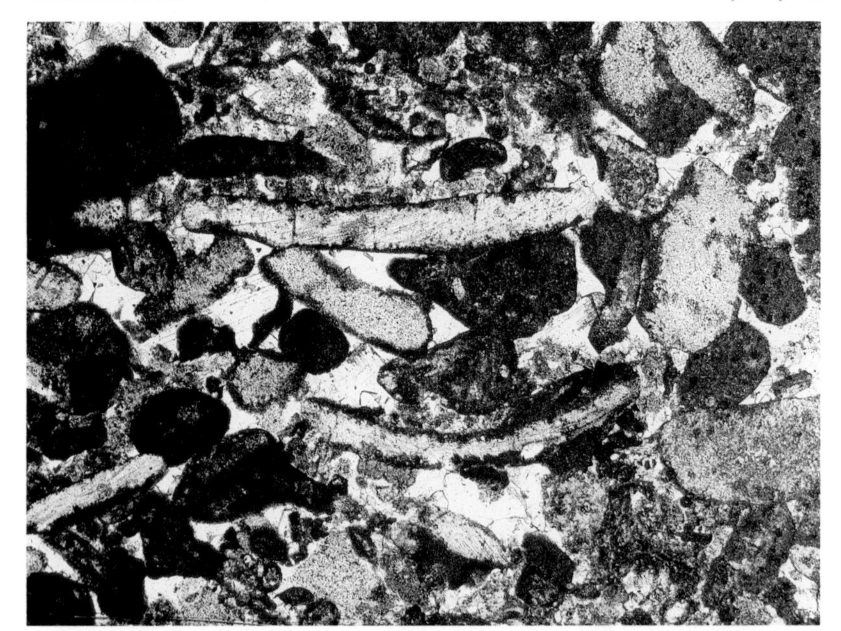

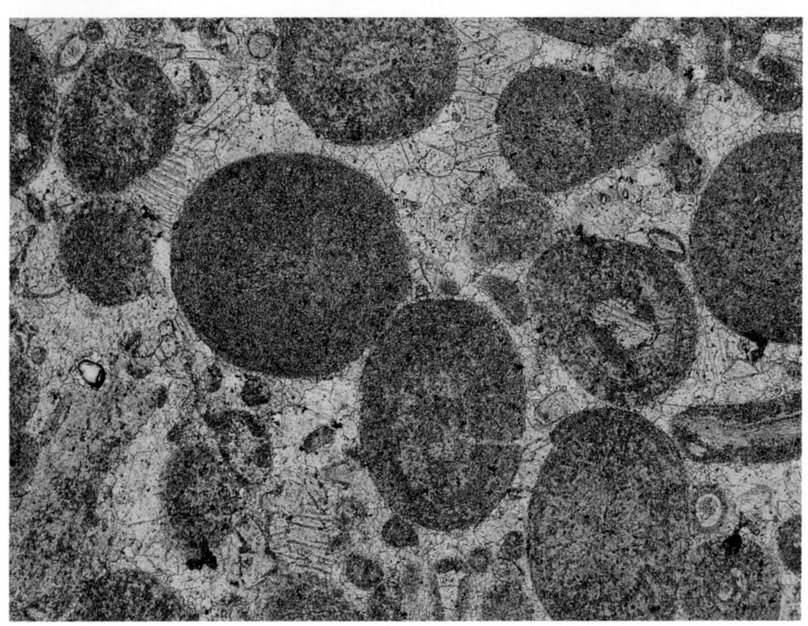

Non-skeletal algae

Micritization

In the shallow marine environment, some non-skeletal blue-green algae may bore into skeletal material. These are called *endolithic algae*. The borings, around 10 μm in diameter, are filled with micrite after the death of the algae. If the process continues, the margin of a shell fragment may become completely replaced by micrite. The process is known as *micritization* and the replaced shell margin as a *micrite envelope*.

123 shows micrite envelopes developed on brachiopod shells (the foliated structure) and echinoderm fragments (the speckled plates). Note the irregularity of the contact between the micrite envelope and the unaltered skeleton. This enables micrite envelopes formed by micritization by blue-green algae to be distinguished from micritic coatings around the *exterior* of skeletal fragments (**77**). Repeated micritization may lead to the production of a grain with no remaining recognizable structure. This would then be called a peloid (**76**). Skeletal algal fragments are often susceptible to this total micritization and it is possible that some of the micritic grains in **123**, with their irregular shape and trace of internal structure, were formed by this process.

124 shows the importance of micrite envelopes in preserving molluscan fragments during diagenesis. The original aragonite molluscan shell has been completely dissolved and the mould, outlined by a thin micrite envelope, was then filled by a sparry calcite cement. Although the sparite is mainly blue-stained ferroan calcite, there are thin zones of pink-stained non-ferroan calcite. This is clearly seen in the shell fragment to the left below the centre.

Allochems other than bioclasts may become micritized. **125** shows a number of grains with varying degrees of preservation of radial and concentric ooid textures (p. 35). It is possible that the texture was partially lost by micritization, although it might also have been lost during inversion of an original aragonite ooid to calcite (neomorphism, p. 60).

123: Unstained thin section, Woo Dale Limestone, Lower Carboniferous, Peak Forest, Derbyshire, England; magnification × 25, PPL.
124: Stained thin section, Inferior Oolite, Middle Jurassic, Leckhampton Hill, Gloucestershire, England; magnification × 12, PPL.
125: Stained thin section, Llandyfan Limestone, Lower Carboniferous, Black Mountains, South Wales; magnification × 43, PPL.

Carbonate cements

The morphology and mineralogy of the pore-filling cement crystals in a limestone can yield information about the environment of cementation. Cements precipitated from marine pore-waters close to the sediment -water interface may be aragonite or high magnesium calcite, but in either case they may form small crystals with a high length-to-width ratio. The crystals are aligned at right angles to the surface on which the cement nucleates. On curved surfaces this means that many marine cements display a *radial-fibrous* fabric.

126 shows a section through a coral skeleton (brownish-stained, structure not clearly visible) in which the first generation of cement is *acicular aragonite* showing a radial-fibrous texture. Note the variation in the length of the crystals which gives a very irregular outer margin to this generation of cement. Such a cement, being aragonite, is not likely to be well-preserved in an ancient limestone. If it undergoes neomorphism (p. 60), the overall radial-fibrous fabric may be retained although detail will be lost. In the sample shown, there is a second generation of pink-stained fine sparite infilling pores. This is typical of cement deposited from meteoric waters.

127 shows a limestone in which there are also two cement generations. The first appears as a rim of crystals of equal thickness on all grains (about 2 mm in width in the photograph). Such cements are said to be *isopachous*. The cement exhibits a radial-fibrous fabric although the length-to-width ratios of the crystals are not as great as those in **120**. It may originally have been aragonite, details of the texture having been lost during inversion to calcite, or it may have been a high magnesium calcite marine cement in which the crystals were elongate prisms rather than needles. The final pore fill is an equant sparite, blue-stained and thus ferroan calcite. This latter cement is characteristic of deposition from meteoric waters or from connate waters fairly deep in the subsurface. In order to incorporate ferrous iron into the calcite lattice to produce a ferroan calcite, reducing conditions must exist. If the pore-waters are oxidizing, any ferrous iron present is rapidly oxidized to ferric iron and precipitated as iron hydroxide. Reducing conditions are more likely to occur at depth than near the surface. Other coarse ferroan calcite cements are seen in **80**, **87**, **90** and **124**.

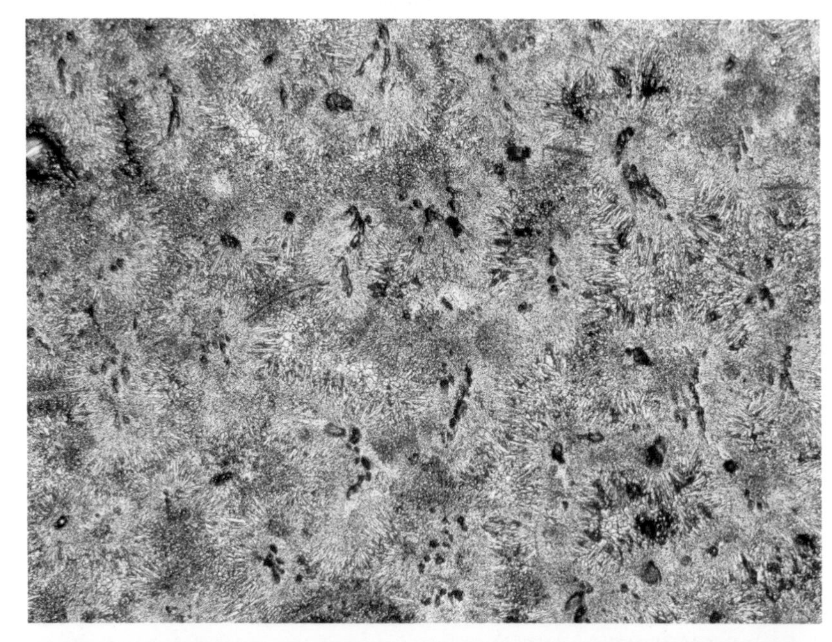

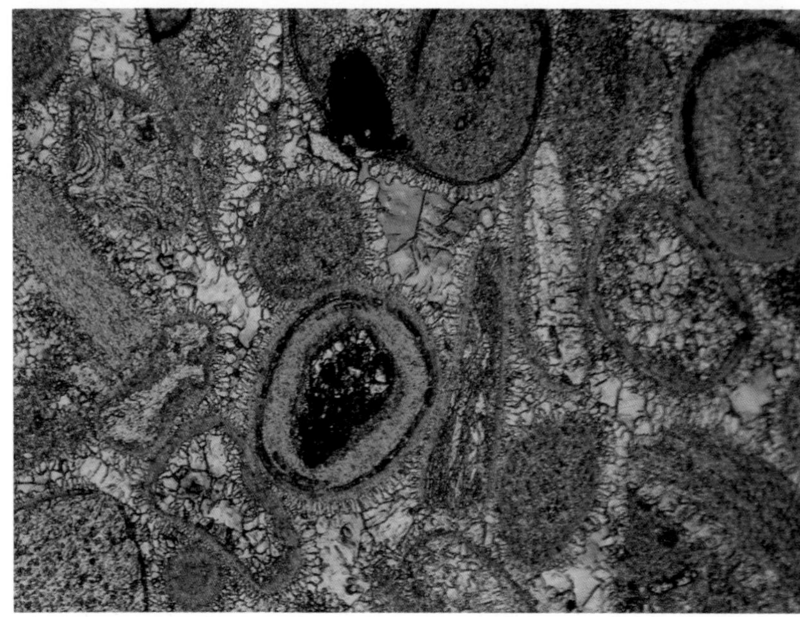

126: Stained acetate peel, Quaternary, Mombasa, Kenya; magnification × 70, PPL.
127: Stained acetate peel, Ouanamane Formation, Middle Jurassic, Western High Atlas, Morocco; magnification × 122, PPL.

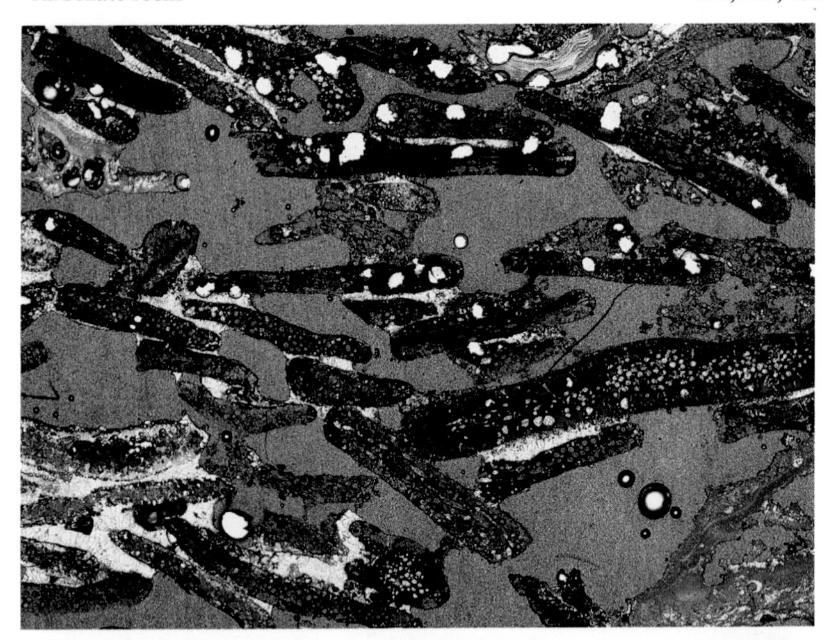

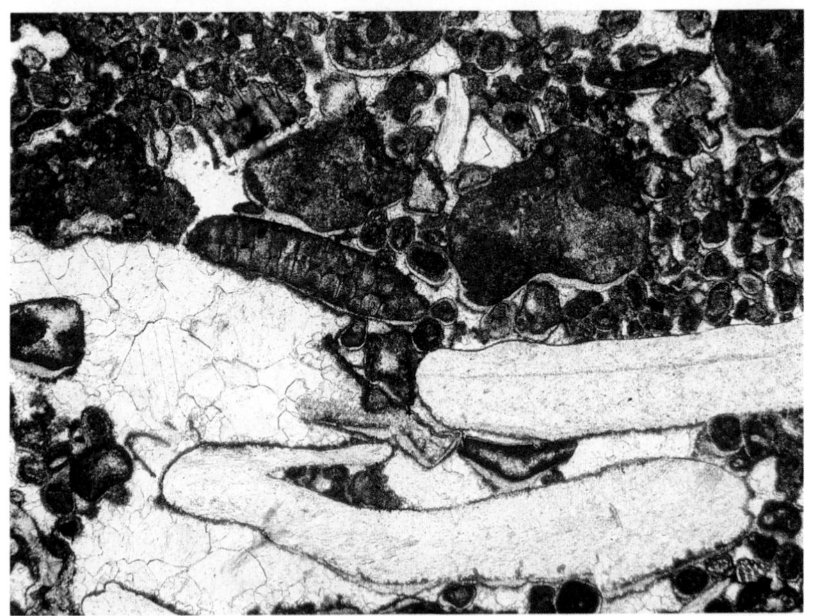

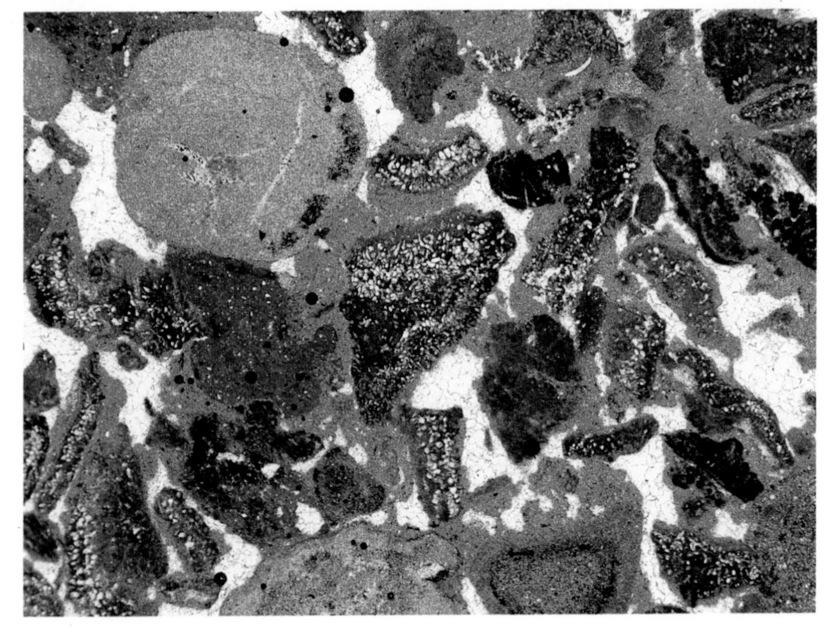

Carbonate cements

(continued)

An early phase of cementation may occur in the vadose zone (above the water table), where pores in the sediment are not completely water-filled. Water, and hence calcite cement, occur around grain contacts in the form of a meniscus.

128 shows a sediment largely made up of segments of the codiacean alga *Halimeda* (**112**). The rock is highly porous, and although impregnated (the brownish-grey background material is the impregnating resin), it has been difficult to take a peel and hence there are numerous air bubbles. The algal segments have been cemented by a small volume of pink-stained fine calcite sparite at grain contacts. This is characteristic of cementation from meteoric waters in the vadose zone. Note the meniscus effect leading to the rounding of pore spaces, well seen to the left of centre of the photograph.

Another feature which can occur in the vadose zone is a *dripstone* or *microstalactitic* cement. In this case water droplets and hence cements are concentrated on the undersurfaces of grains. **129** shows a sediment in which the first generation of cement occurs only on the lower surfaces of some grains. In the photograph it is very pale, brownish-coloured and never more than a millimetre thick. Vadose cements can form from marine pore waters in the intertidal and supratidal zones as well as from meteoric waters. In the former case the cement will have a radial-fibrous fabric. In the example here, the cement is too fine-grained for its fabric to be resolved at the magnification shown. A later generation of coarse sparite fills the pores.

Cements, especially those deposited in a marine environment, may be micrite. In ancient limestones where pore-spaces are completely filled, it is difficult to distinguish micrite cements, which have nucleated on grain surfaces and grown outwards to fill or partially fill pore-spaces, from carbonate mud sediment deposited with the grains. **130** illustrates a sediment comprising fragments of algae and micrite peloids having a matrix which is a mixture of micrite (greenish-brown) and sparite (colourless). The micrite coats some grains and forms 'bridges' between adjacent grains and it may therefore be a cement. However, it is possible that micrite sediment, deposited along with the grains, became partially lithified and was then subject to erosion by a through flow of water which removed unlithified material.

***128**: Stained acetate peel, Quaternary, Mombasa, Kenya; magnification × 9, PPL.*
***129**: Unstained thin section, Woo Dale Limestone, Lower Carboniferous, Long Dale, Derbyshire, England; magnification × 22, PPL.*
***130**: Unstained thin section, Coal Measures, Upper Carboniferous, Metallic Tileries, Chesterton, Staffordshire, England; magnification × 20, PPL.*

Carbonate cements

(continued)

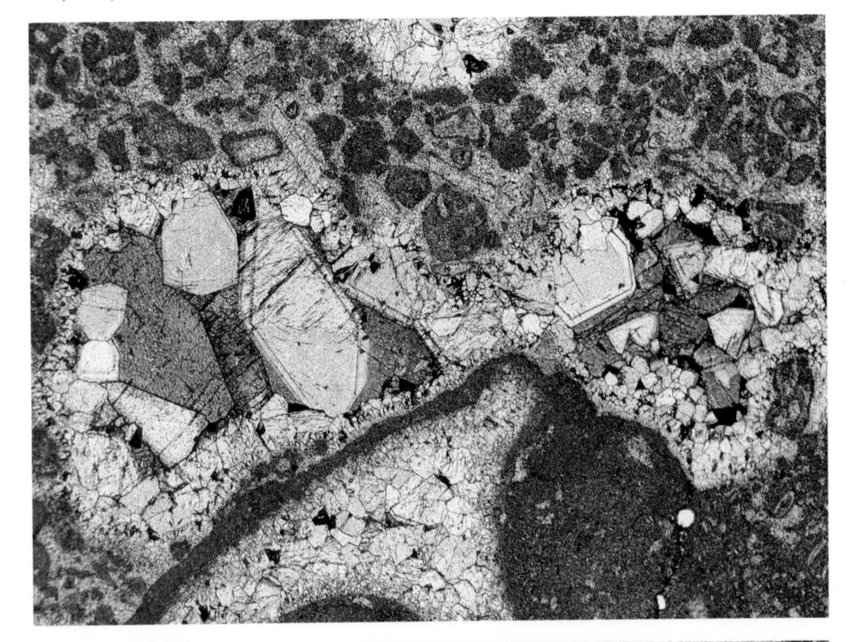

The characteristic texture of a cement precipitated in the meteoric phreatic zone (below the water-table) is one in which the crystals increase in size from the margins of pores towards their centres. This is known as a *drusy mosaic* and results from competitive growth of crystals away from the substrate on which they nucleate. The resulting fabric is one of more or less equidimensional crystals, sometimes known as 'blocky' or 'equant' sparite.

131 illustrates a drusy mosaic in which cement crystals show compositional zoning, the stain picking out changes in the amount of iron in the calcite brought about by changes in composition of the circulating groundwaters. As explained on page 55, ferroan calcite cements are precipitated under reducing conditions. The zoning indicates the position of crystal faces during growth, and shows that the crystals were euhedral at the time, although growth to completely fill the pore spaces has led to the final crystal shapes being anhedral. Crystal boundaries formed by crystals growing together in this way are known as *compromise boundaries*.

Where the component grains of a limestone are composed of a few large crystals, it is often possible to see that cements have been precipitated in optical continuity with the grains on which they nucleate. These are known as *syntaxial overgrowths* or *syntaxial rim cements* and are most easily seen on echinoderm fragments. **132** and **133** show a sediment in which the cement is composed entirely of syntaxial overgrowths on crinoid plates. The crinoids can be identified by their speckled appearance whereas the cement is clear. The syntaxial nature of the cement is shown in PPL by the cleavage passing through both bioclast and cement. In XPL, the uniform extinction colour of both crinoid and overgrowth can be seen. Fragments of fenestrate bryozoans are abundant in the sample.

***131**: Stained acetate peel, Woo Dale Limestone, Lower Carboniferous, Wolfscote Dale, Staffordshire, England; magnification × 22, PPL.*
***132** and **133**: Stained thin section, Eyam Limestone, Lower Carboniferous, Once-a-week Quarry, Derbyshire, England; magnification × 27; **132** PPL, **133** XPL.*

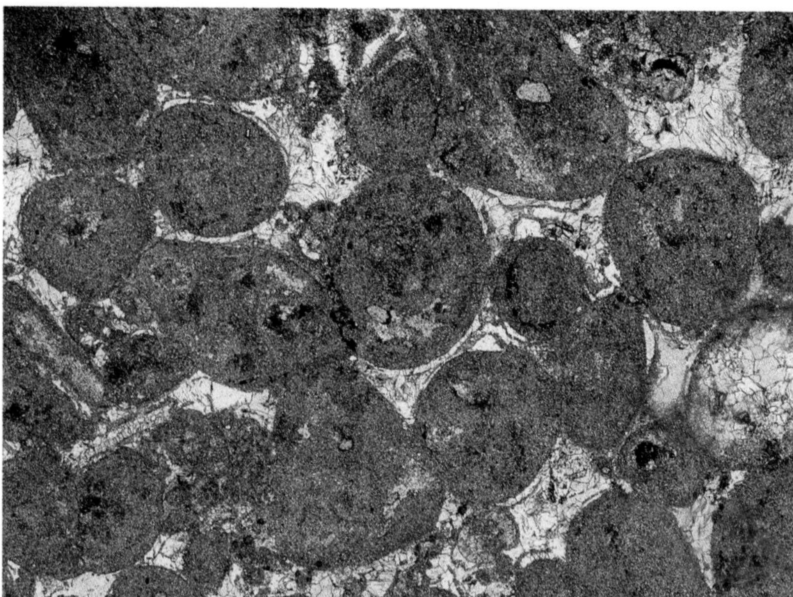

Compaction

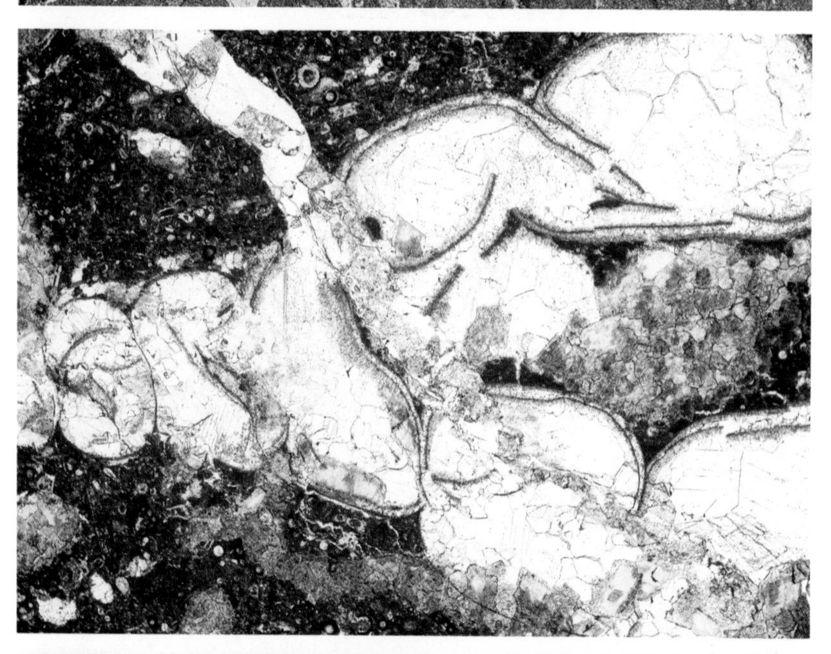

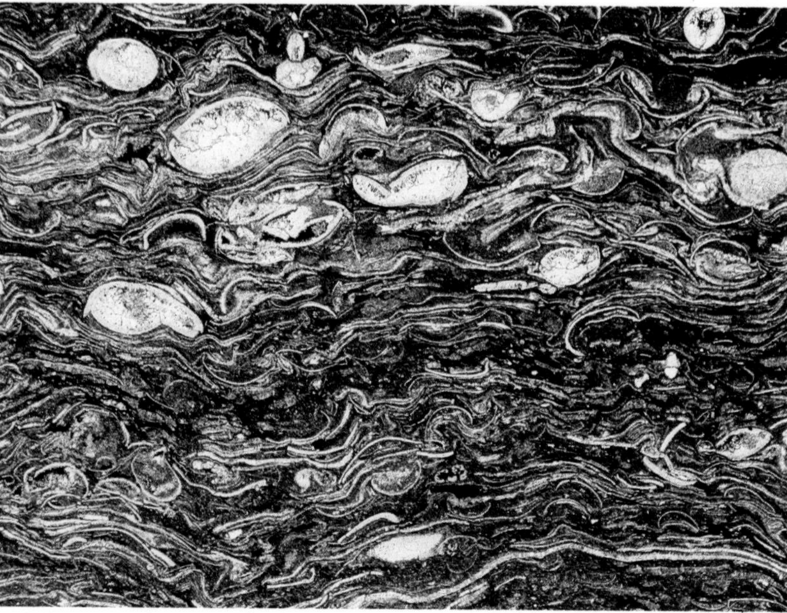

Apart from cementation, the major process leading to porosity reduction in sediments is compaction. Early stages of compaction in uncemented sediments involve the readjustment of loose grain fabrics to fit more tightly together, the fracture of delicate shells, the squashing of soft grains, and the dewatering of carbonate mud.

134 shows a peloidal limestone in which either the outer layers of the peloids, or a very thin early generation of cement, has flaked off during compaction. The micritic grains must have been rigid or compaction would have resulted in their deformation. Compaction was followed by the precipitation of a coarse sparite cement which 'healed' the fractures caused by the flaking off of the rinds of the grains.

135 shows a cross-section of a gastropod preserved as a cast. The inner wall of the organism is marked by a micrite envelope and a thin generation of early cement (see for example the chambers in the upper part of the photograph). The wall of the shell has been fractured and some fragments disoriented during compaction. Both micrite envelope and early cement are fractured and the fractures then healed by a coarse sparite cement. Thus after deposition, the mollusc was micritized and then cemented by a thin early generation of fine carbonate. Then the aragonite wall was dissolved and fracturing occurred, before the rock was finally cemented. The sample also shows a vein running from top left to bottom right of photograph and brown-coloured replacement dolomite crystals are scattered throughout the sediment.

136 illustrates a highly compacted bioclastic sediment, consisting of complete two-valved ostracods as well as single ostracod valves and long, thin bivalve fragments. Most fragments are aligned parallel to the bedding but some still show folding and fracturing (e.g. upper left). The complete ostracods have withstood considerable pressure but most eventually fractured.

134: Stained acetate peel, Red Hill Oolite, Lower Carboniferous, Cumbria, England; magnification × 31, PPL.
135: Unstained thin section, Woo Dale Limestone, Lower Carboniferous, Derbyshire, England; magnification × 14, PPL.
136: Unstained thin section, Coal Measures, Upper Carboniferous, Cobridge Brickworks, Hanley, Staffordshire, England; magnification × 19, PPL.

Pressure-solution and deformation

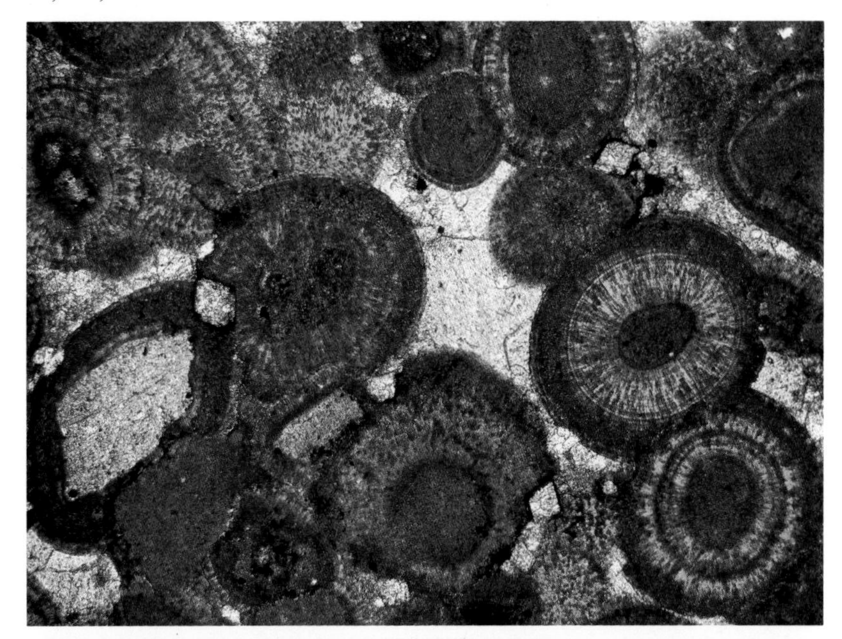

Pressure-solution is the process whereby a sediment, because it is under load, is subject to selective dissolution. In limestones it is normally calcium carbonate that is dissolved and any less soluble material such as clay and quartz is concentrated along seams.

137 illustrates a case of grain-to-grain pressure-solution. Before the pores of a rock are filled by cement, stress is concentrated at the points where the grains meet and part of one or both the grains dissolves. In the example, ooids have undergone solution. The later cement is a mauve-stained, slightly ferroan sparite. Note the small rhombic areas of fine calcite spar (e.g. midway up, half-way between centre and left-hand edge). These are calcite pseudomorphs after dolomite (dedolomite, p. 74).

138 shows a limestone which has undergone pressure-solution to such an extent that most grain boundaries have been modified and the rock is pervaded by thin dark seams. Many of these have the fine saw-tooth appearance characteristic of stylolites. This type of pervasive pressure-solution is known as sutured-seam solution.

139 shows a limestone which has been subjected to some stress. Speckled echinoderm plates are recognizable, together with syntaxial overgrowths. Most of the calcite crystals are twinned, a feature which may develop as a result of pressure, and the twin planes can be seen to be slightly bent.

137: Stained thin section, Upper Jurassic, Cap Rhir, Morocco; magnification × 52, PPL.
138: Stained acetate peel, Woo Dale Limestone, Lower Carboniferous, Long Dale, Derbyshire, England; magnification × 31, PPL.
139: Stained thin section, Torquay Limestone, Devonian, Hope's Nose, Devon, England; magnification × 31, PPL.

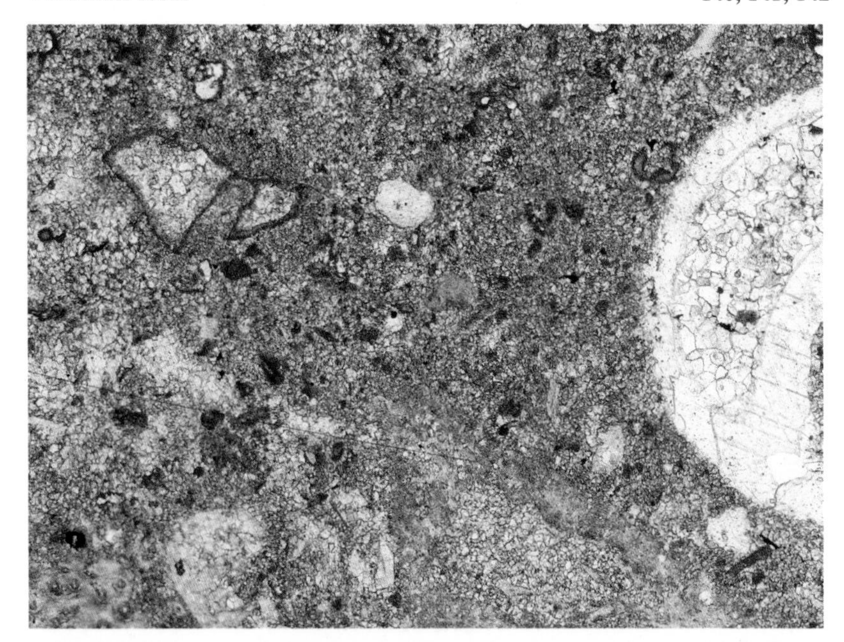

Neomorphism

Microspar, Pseudospar

The term *neomorphism* refers to all transformations between a mineral and the same mineral, or another of the same general composition. During diagenesis, aragonite components of limestone may be transformed to calcite without the development of significant porosity. There is usually an accompanying increase in grain size (aggrading neomorphism). In particular the micritic material of limestones may be altered to coarser calcite. The terms *microspar* and *pseudospar* are used for crystal mosaics of neomorphic origin having a mean size of between 4 and 10 μm and > 10 μm respectively. It is not always possible to differentiate between neomorphic fabrics and fine spar cements, or sediments composed of primary silt-sized particles. As a guide, neomorphic spar generally has irregular crystal boundaries and patchy grain size distribution, often with relicts of micrite and 'floating' skeletal grains.

140 shows a limestone in which the matrix is fine pseudospar. It is cloudy and contrasts with the coarse clear mosaic replacing the wall and infilling the chamber of the mollusc on the right of the photograph. Its grain size varies patchily and it is thus likely to be neomorphic, having originally been a micritic sediment.

141 shows a very fine-grained limestone (note magnification) composed almost entirely of calcium carbonate grains of microspar size. There appear to be no micritic relicts and this fabric may be a primary one as a result of deposition of carbonate mud of fine silt-sized particles, rather than a product of neomorphism of micrite.

142 illustrates a limestone with a few dolomite rhombs (dark-coloured) in a 'matrix' of pseudospar with patches of microspar and micrite. Note how crystal size and shape vary irregularly throughout the mosaic. This is characteristic of a neomorphic fabric.

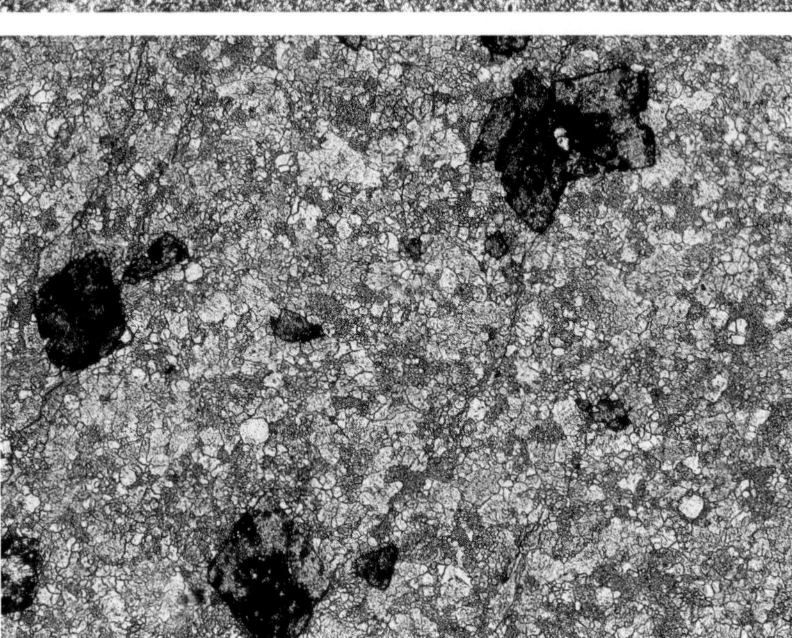

140: Stained thin section, Carboniferous Limestone, Llangollen, North Wales; magnification × 43, PPL.
141: Stained acetate peel, Blue Lias, Lower Jurassic, Lavernock Point, South Wales; magnification × 72, PPL.
142: Stained acetate peel, Woo Dale Limestone, Lower Carboniferous, Woo Dale, Derbyshire, England; magnification × 43, PPL.
A neomorphic fabric is also shown in **161**.

Neomorphism

Bioclasts

Most aragonite bioclasts are preserved as calcite casts with no trace of the original microstructure of the wall (see p. 40). Occasionally, however, aragonite bioclasts invert to calcite *in situ*. This is a form of neomorphism.

143 illustrates parts of the shells of bivalves which have been subject to neomorphism. The shells consist of a blue-stained ferroan calcite sparry mosaic, but there are lines of inclusions cutting across crystal boundaries and indicating the original foliated structure of the shell. Many crystals are also brown-coloured because of their inclusion content. The sediment between the shells is muddy and contains abundant quartz (unstained).

144 and **145** show sections through a colonial scleractinian coral, originally composed of aragonite and now calcite. **144** shows the coral at low magnification. The coral walls and septa are inclusion-rich, non-ferroan calcite of which the detailed fabric is unclear. Pore-space is filled with an inclusion-free sparite cement, very pale mauve-stained and hence slightly ferroan. **145** shows the colony at higher magnification in a section which has been ground slightly thinner than is usual. The coral walls comprise an irregular mosaic of crystals of varying grain sizes and shapes which is neither the original microstructure nor a drusy mosaic, but a product of neomorphism. Note that some crystal boundaries cut across from pore-filling sparite to coral septa and thus some crystals are partly neomorphic and partly cement.

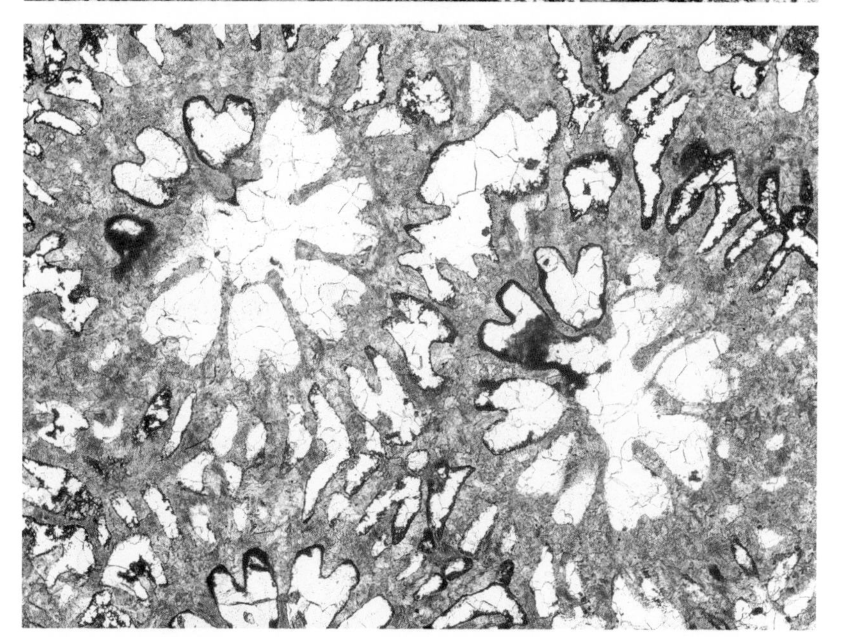

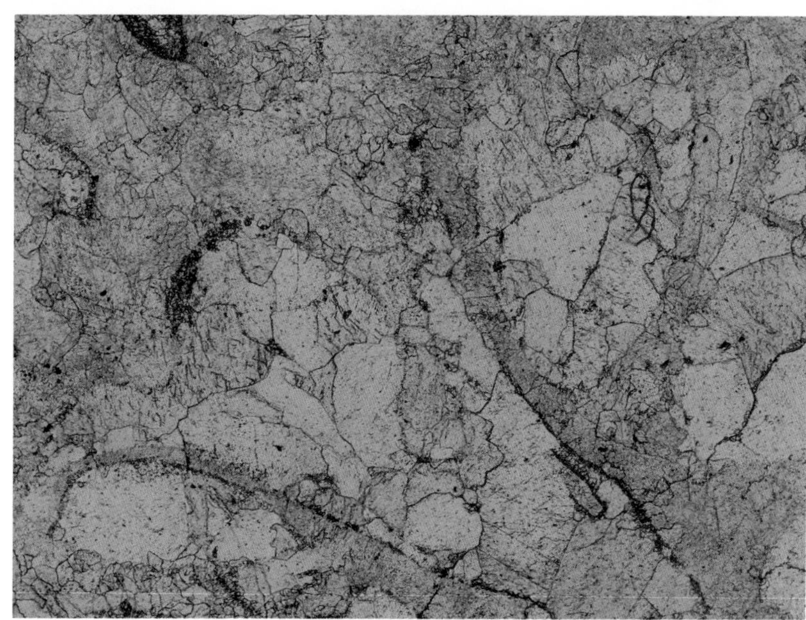

***143**: Stained thin section, Wealden, Lower Cretaceous, southern England; magnification ×21, PPL.*
***144** and **145**: Stained thin sections, Iggui el Behar Formation, Upper Jurassic, Imouzzer des Ida-ou-Tanane, Western High Atlas, Morocco; magnification **144** ×12; **145** ×43, PPL.*

Limestone Classification

Two of the most popular limestone classifications are those of Folk (1959, 1962) and Dunham (1962). These are summarized in Tables 3 and 4 and Fig. E.

Table 3. Classification of limestones according to Dunham (1962) Rock names are in capital letters

Original components not organically bound together during deposition				Components organically bound during deposition
contains carbonate mud			no carbonate mud	
mud-supported		grain-supported		
< 10% allochems	> 10% allochems			
MUDSTONE	WACKESTONE	PACKSTONE	GRAINSTONE	BOUNDSTONE

Table 4 Classification of limestones based on the scheme of Folk (1959, 1962) Rock names are in capital letters

volumetric allochem composition			> 10% allochems		< 10% allochems		
			Sparry calcite > Micrite	Micrite > Sparry calcite	1–10% allochems	< 1% allochems	Undisturbed reef and bioherm rocks
> 25% Intraclasts			INTRASPARITE	INTRAMICRITE	Most abundant allochems: Intraclasts INTRACLAST-BEARING MICRITE	MICRITE, or if sparry patches present DISMICRITE	BIOLITHITE
< 25% Intraclasts	> 25% Ooids		OOSPARITE	OOMICRITE	Ooids OOID-BEARING MICRITE		
< 25% Intraclasts	< 25% Ooids; Volume ratio, bioclasts: peloids	> 3:1	BIOSPARITE	BIOMICRITE	Bioclasts FOSSILIFEROUS MICRITE		
< 25% Intraclasts	< 25% Ooids; Volume ratio, bioclasts: peloids	3:1 to 1:3	BIOPELSPARITE	BIOPELMICRITE			
< 25% Intraclasts	< 25% Ooids; Volume ratio, bioclasts: peloids	< 1:3	PELSPARITE	PELMICRITE	Peloids PELOID-BEARING MICRITE		

Over ⅔ micrite matrix				Subequal spar & micrite	Over ⅔ spar cement		
0–1% Allochems	1–10% Allochems	10–50% Allochems	Over 50% Allochems		Sorting poor	Sorting good	Rounded & abraded
Micrite & dismicrite	Fossiliferous micrite	Sparse biomicrite	Packed biomicrite	Poorly-washed biosparite	Unsorted biosparite	Sorted biosparite	Rounded biosparite

Micrite matrix   Sparry calcite cement

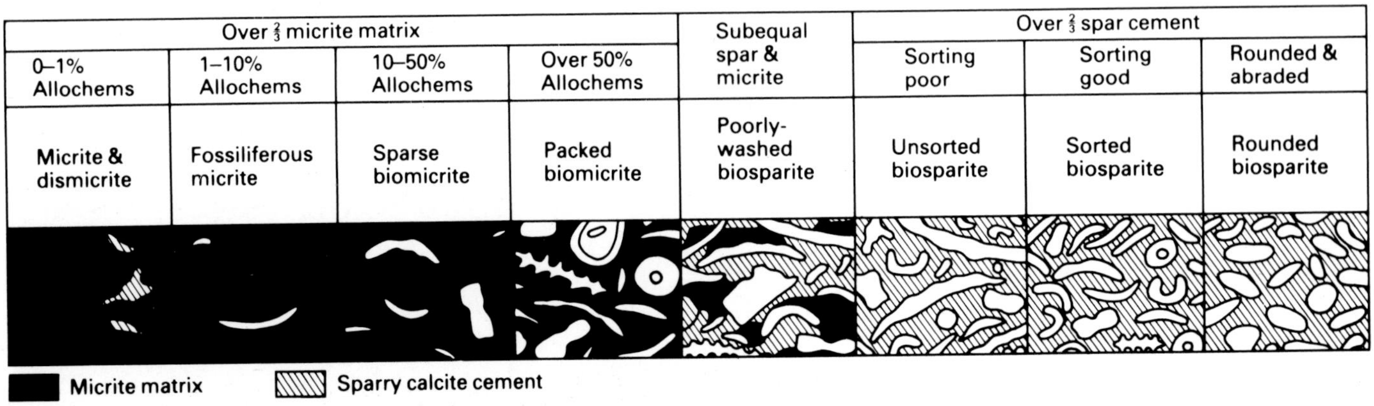

Fig. E The range in textures shown by carbonate rocks, illustrated using the rock names of the Folk classification (after Folk, 1959)

Limestone classification

(continued)

146 illustrates a grainstone. The rock is grain-supported with a spar cement. The sediment is loosely-packed, suggesting that cementation occurred before significant compaction. The allochems are a mixture of ooids (some are superficial ooids, see p. 35) and bioclasts. It is therefore an *oosparite* according to Folk. Since the allochems are rounded it would be a *rounded oosparite*, using Folk's textural spectrum. **147** shows a packstone. The rock shows two sizes of grains, having large and small peloids. The former have a trace of oolitic structure in places and may be micritized ooids (p. 54). The latter are probably faecal pellets. The sediment contains some ferroan calcite cement but also much carbonate mud sediment in the matrix. It is nevertheless grain-supported and thus a packstone. According to Folk's classification it is a *poorly-washed oosparite*.

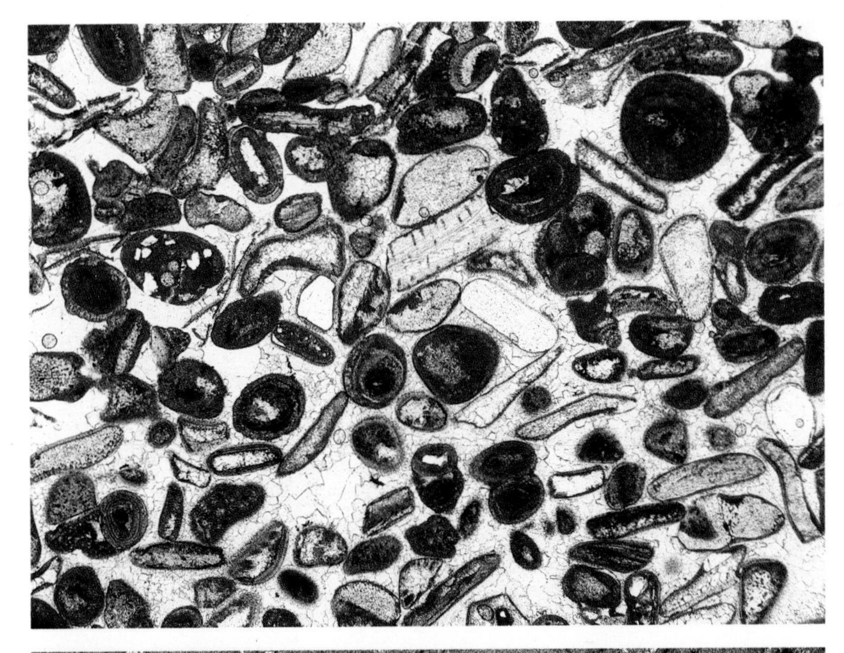

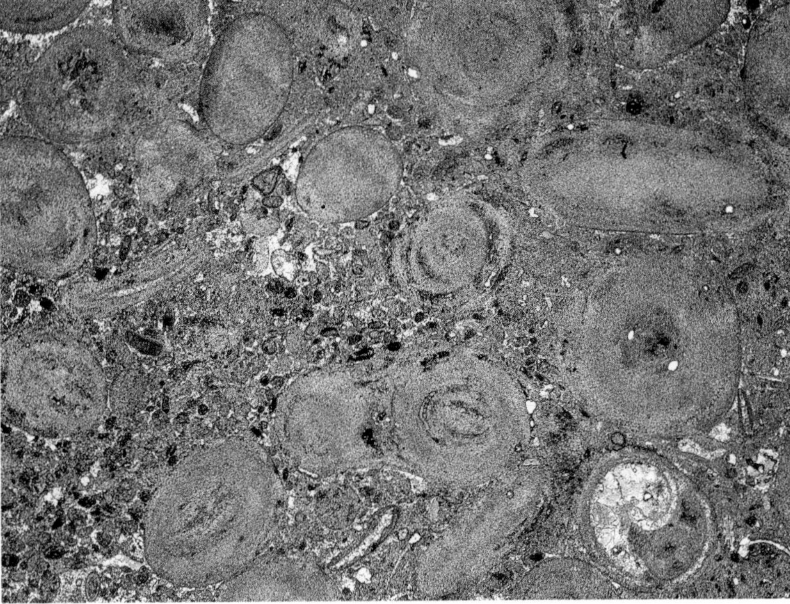

146: Unstained thin section, Jurassic, unknown locality, England; magnification × 23, PPL.
147: Stained acetate peel, Inferior Oolite, Middle Jurassic, Cooper's Hill, Gloucestershire, England; magnification × 13, PPL.
Other grainstones are shown (Folk classification in brackets) in **73** *(oosparite),* **74** *(oosparite),* **75** *(sorted pelsparite),* **77** *(unsorted intrasparite),* **87** *(unsorted biosparite) and* **124** *(unsorted biosparite).*
Other packstones are shown in **72** *(poorly-washed oosparite),* **79** *(packed intramicrite)* **96** *(poorly-washed biosparite) and* **115** *(packed biomicrite).*

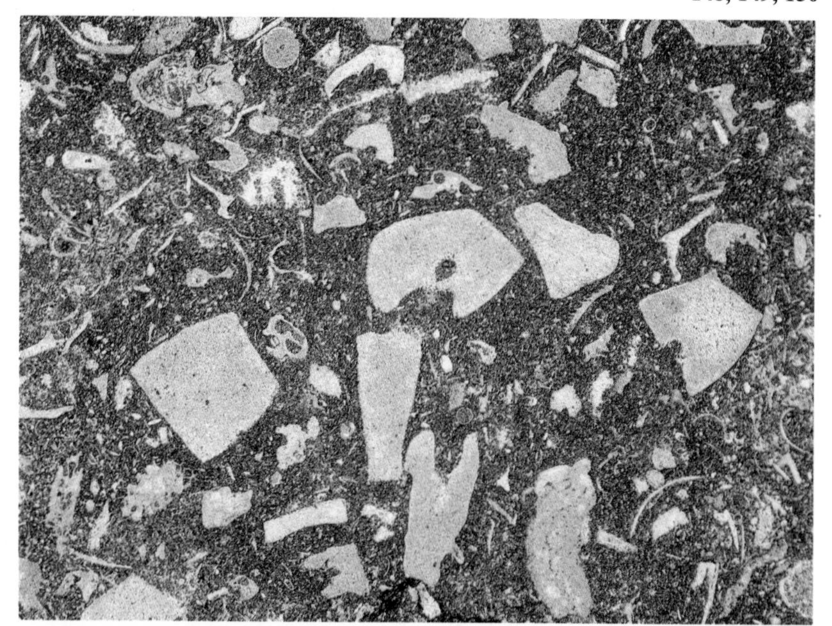

Limestone classification

(continued)

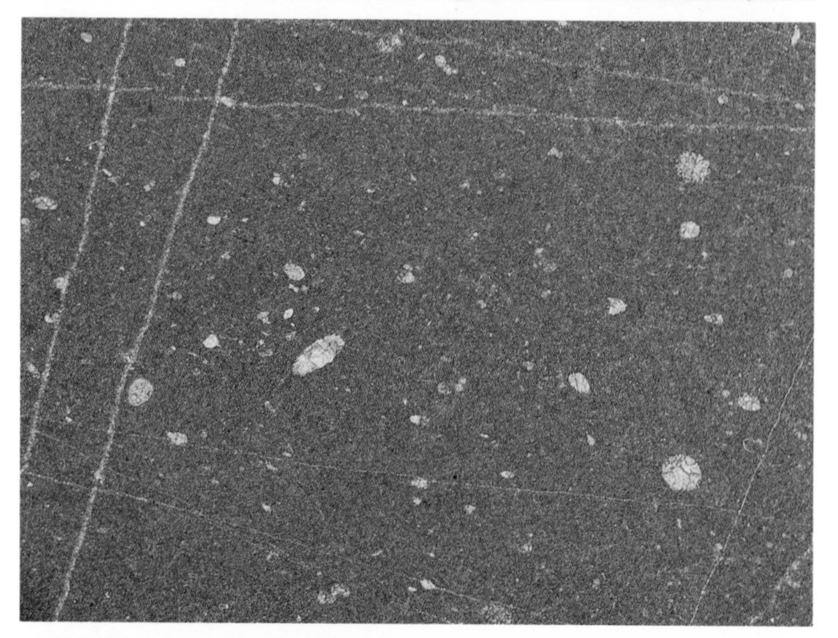

148 shows a wackestone. The grains are bioclasts, mainly echinoderm plates with some bryozoans (e.g. lower left part). These grains are supported by a matrix of carbonate mud in which many small particles are visible at this magnification.

149 shows a mudstone, being a matrix-supported limestone with less than 10% allochems. In this case the allochems are microfossils – foraminifera and calcite casts of radiolaria. The sediment is cut by thin veins of pale blue-stained ferroan calcite. This sample is a *fossiliferous micrite* according to Folk's classification.

A boundstone is a limestone in which sediment is bound together by organisms, such as occurs in many reefs. Textures are often more clearly visible at hand-specimen scale. **150** shows a thin section of a reef limestone comprising growths of a number of problematic organisms (probably algae or foraminifera) which have encrusted one another while incorporating fine-grained sediment into the rock framework.

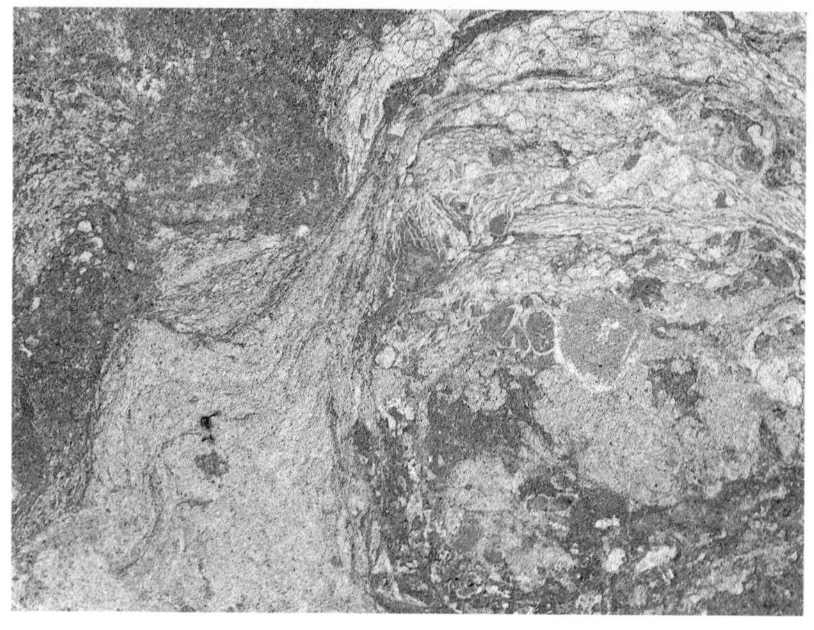

148: Stained acetate peel, Wenlock Limestone, Silurian Shropshire, England; magnification × 11, PPL.
149: Stained thin section, Upper Cretaceous, Pindos Zone, Greece; magnification × 43, PPL.
150: Stained thin section, Red Hill Oolite, Lower Carboniferous, Elliscales Quarry, Dalton-in-Furness, Cumbria, England; magnification × 12, PPL.
*Other wackestones are shown (Folk classification in brackets), in **105** (biomicrite) and **156** (biomicrite).*

Limestone Porosity

Any description of a limestone should include an evaluation of the amount and type of porosity in the sediment. Porosity may be primary, having been present in the rock since deposition, or secondary, having developed as a result of diagenesis. A classification of porosity types is shown in Fig. F. The terminology of porosity types illustrated here with limestones, is also applicable to sandstones.

Fabric selective

Interparticle

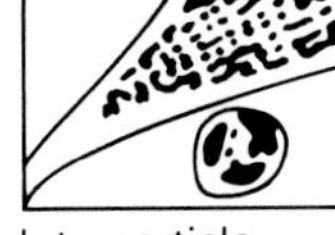
Intraparticle

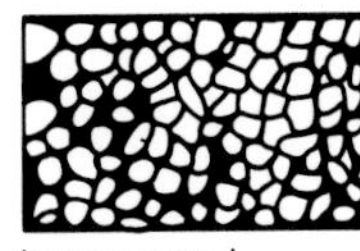
Intercrystal

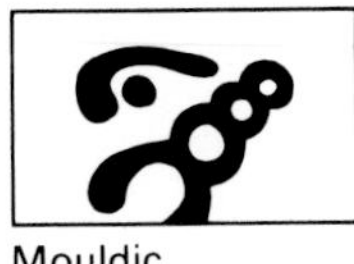
Mouldic

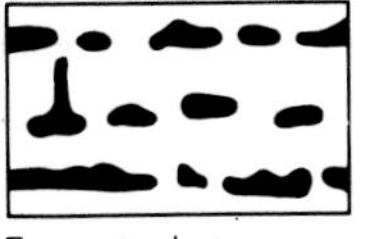
Fenestral

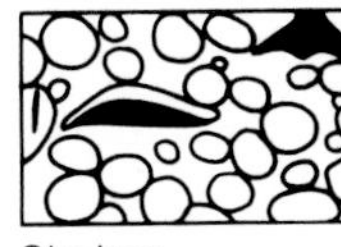
Shelter

Growth framework

Not fabric selective

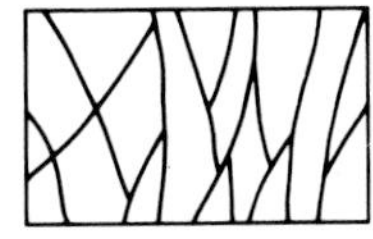
Fracture

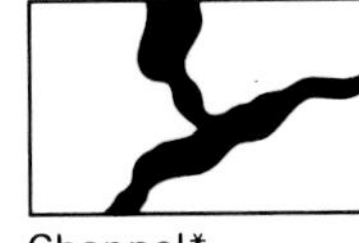
Channel*

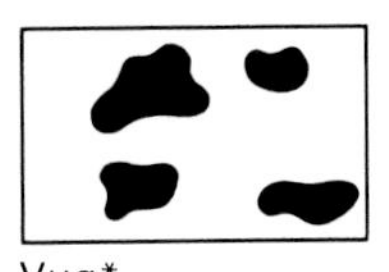
Vug*

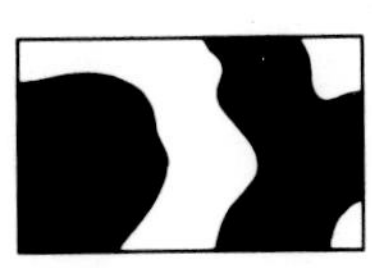
Cavern*

* Cavern applies to man-sized or larger pores of channel or vug shapes

Fabric selective or not

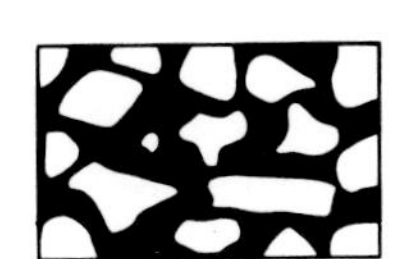
Breccia

Boring

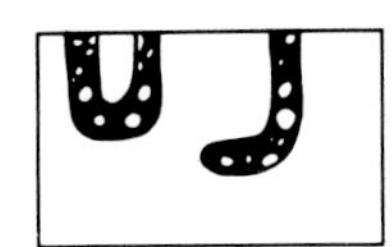
Burrow

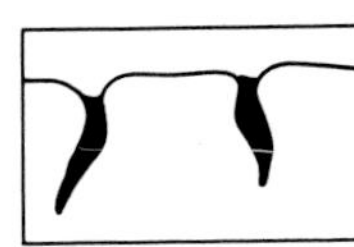
Shrinkage

Fig. F Basic porosity types in sediments. Pores shaded black (after Choquette and Pray, 1970)

Limestone porosity

(continued)

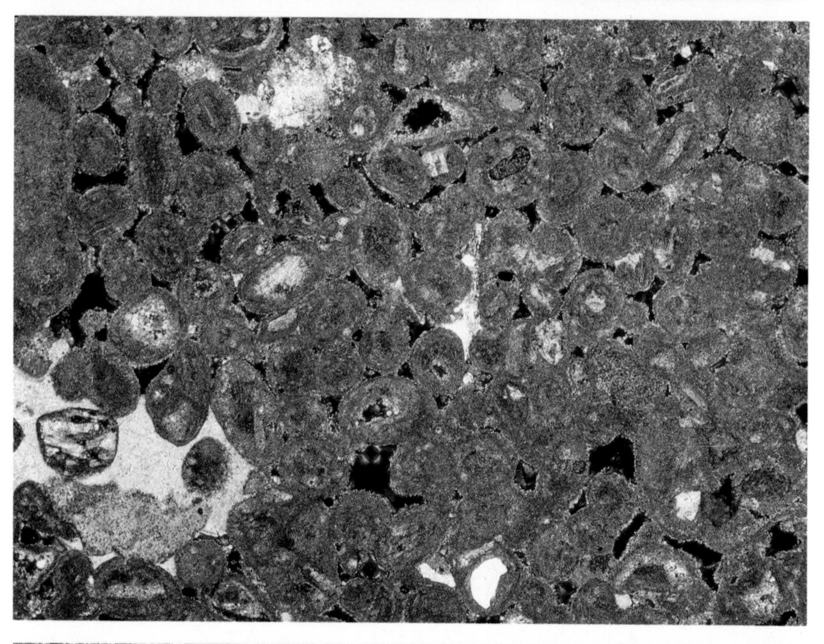

151 and **152** show an oolitic/peloidal sediment in which much of the depositional space between grains is unfilled by sediment or cement. The rock is said to show primary *intergranular* porosity. When deposited, such a sediment may have had as much as 50% pore-space. This has been reduced by compaction and by the introduction of some cement. Two types of cement are present – a fine spar, forming coatings on most grains (about $\frac{1}{2}$ mm thick at this magnification and best seen in XPL) and syntaxial overgrowths on echinoderms (lower left). Although localized, the latter are volumetrically more significant.

A common type of secondary porosity is *mouldic* porosity, usually formed by the dissolution of aragonite bioclasts. **153** shows a sediment having primary intergranular and secondary mouldic porosity. Thin micrite envelopes have supported the shell moulds, although that supporting the fragment seen in the lower part of the photograph has partly collapsed.

The bluish-grey interference colours seen in the intergranular pores and the shell moulds of **152** and **153** are caused by strain in the mounting medium.

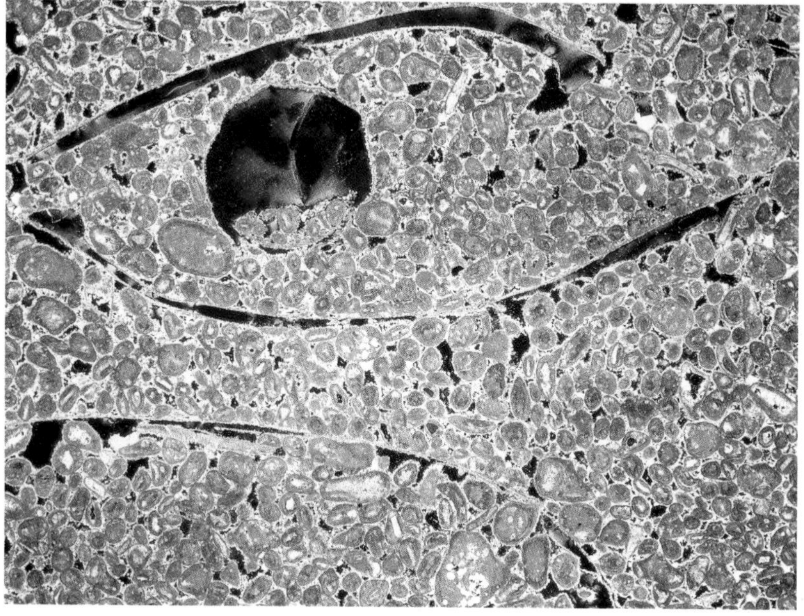

151 and 152: Stained thin section, Portland Stone, Upper Jurassic, Dorset, England; magnification × 27, ***151*** *PPL;* ***152*** *XPL.*

153: Stained thin section, Portland Stone, Upper Jurassic, Dorset, England; magnification × 11, XPL.

Limestone porosity

(continued)

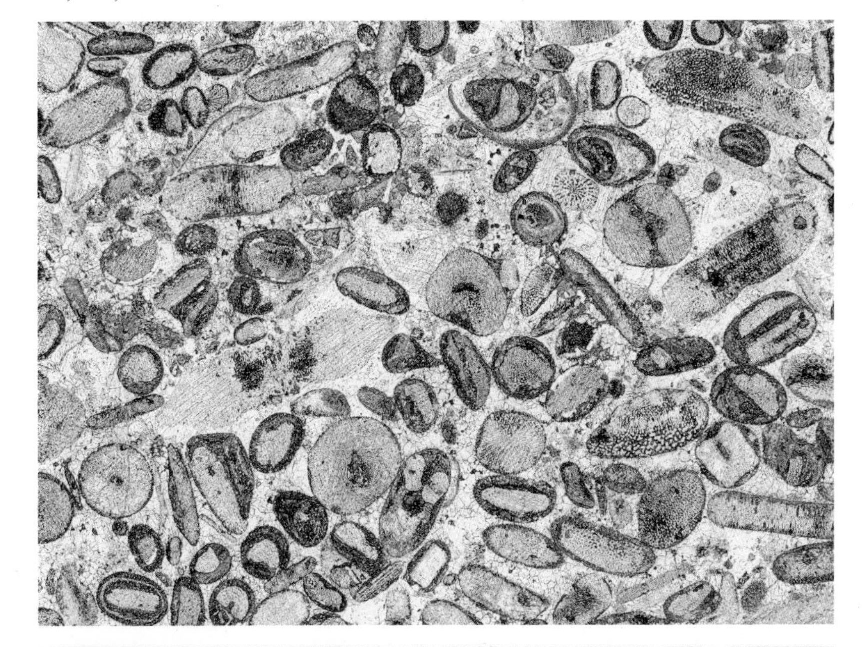

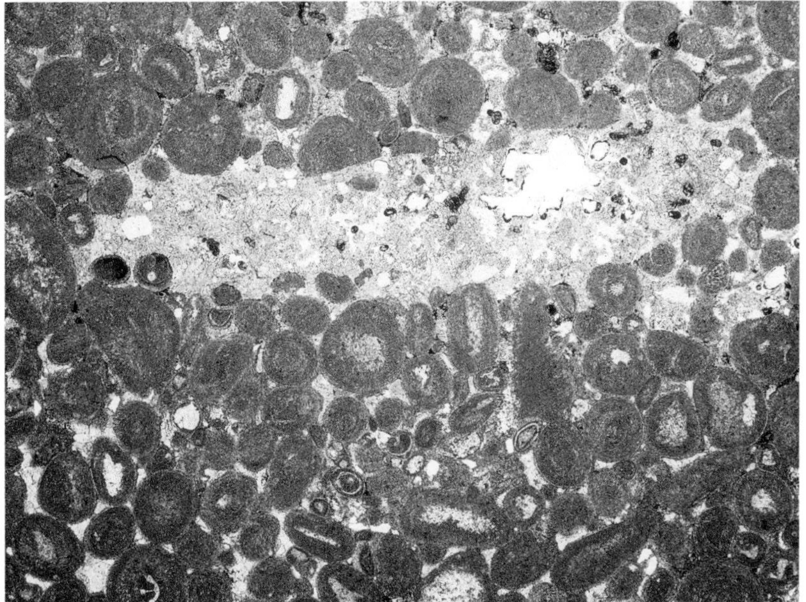

154 shows a limestone composed mainly of echinoderm fragments in a pink-stained, non-ferroan, calcite sparite cement. However, a number of grains comprising a small echinoderm fragment nucleus, surrounded by a zone of blue-stained ferroan calcite cement, are also present. This cement is interpreted as a late infilling of pore-space formed by the dissolution of an aragonite coating to the echinoderm fragments. Such a coating is likely to have been oolitic and after solution the sediment would have exhibited *oomouldic* porosity.

Porosity may develop as a result of the burrowing and boring activities of organisms. **155** Shows a section through a boring made by an organism in an oolitic sediment. Note that grains are truncated around the margins of the boring, indicating that the sediment was lithified when the organism was at work and hence the structure is a boring rather than a burrow. The boring is infilled with a ferroan calcite cement, some of which has been lost during the making of the section.

Shelter porosity occurs below curved shell fragments which are preserved in a convex-up position. **156** shows bivalve fragments in a carbonate mud sediment. Those preserved in a convex-up position, including the large fragment extending right across the field of view, have areas of sparite cement below them which was precipitated during the infilling of shelter cavities. Sediment was unable to fill the cavities because of the 'umbrella' effect of the shell.

154: Stained acetate peel, Oolite Group, Lower Carboniferous, Daren Cilau, Llangattock, South Wales; magnification × 15, PPL.
155: Stained thin section, Inferior Oolite, Middle Jurassic, Cooper's Hill, Gloucestershire, England; magnification × 16, PPL.
156: Stained thin section, Lower Carboniferous, Arbigland, Dumfries, Scotland; magnification × 16, PPL.

Limestone porosity

(continued)

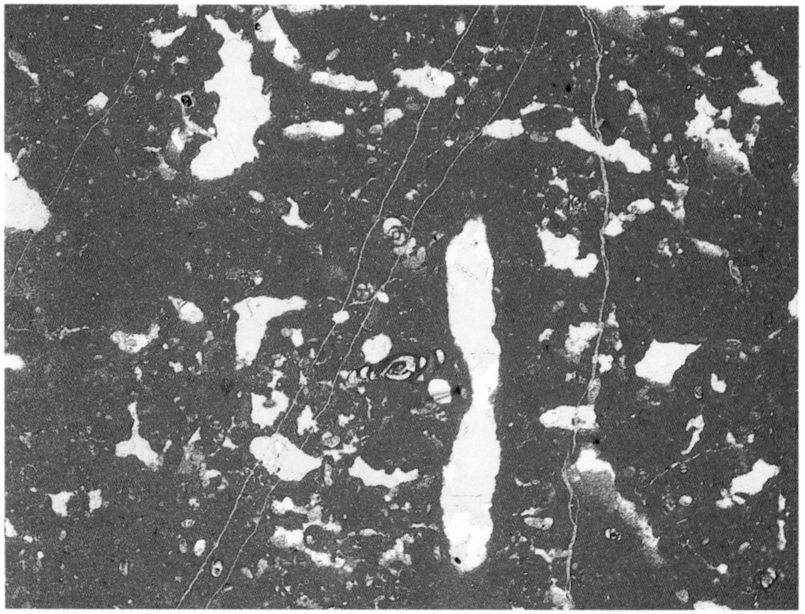

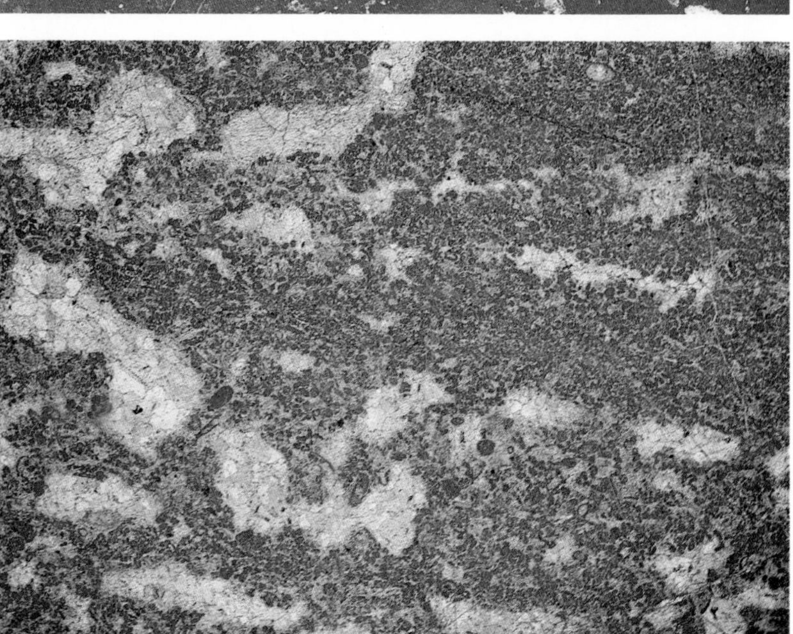

Fenestrae is the name used for pores in a carbonate sediment which are larger than grain-supported spaces. They usually become infilled with internal sediment or cement, or a combination of the two. Fenestrae can be different shapes and sizes depending on their mode of origin.

157 shows spar-filled fenestrae in a micrite. Most are irregular in shape and probably formed as a result of the entrapment of fluid in a sediment during desiccation, although the elongated fenestra in the centre may have been a burrow. Fenestrae of this type are sometimes called *birds-eye structures*. The sediment contains a few micrite-walled foraminifera. Fenestral micrites were called *dismicrite* by Folk (see Table 4).

158 shows fenestrae in a fine pellet grainstone. They show a tendency to be elongate parallel to the bedding. Fenestrae of this type are known as *laminoid fenestrae* and may form from the decay of organic matter associated with algal stromatolites (p. 53).

157: Stained thin section, Lower Jurassic, Central High Atlas, Morocco; magnification × 14, PPL.
158: Stained acetate peel, Woo Dale Limestone, Lower Carboniferous, Derbyshire, England; magnification × 7, PPL.

Limestone porosity

(continued)

Pore-space in limestones may be filled with sediment as well as cement. Sediment partially infilling cavities, particularly in fossils or fenestrae, will indicate the horizontal plane at the time of its deposition. Such sediment infills are known as geopetal infills. **159** shows geopetal sediment within a gastropod. On deposition the gastropod would have had a primary porosity within its chambers (*intragranular porosity*). This was partially infilled by micritic sediment and the cavity finally filled by ferroan calcite cement. Inclusions within the shell wall of the gastropod and surrounding bioclasts suggest that they inverted to calcite during neomorphism (p. 61), rather than being cement-filled casts.

Some pore-spaces have hydrocarbons within them or have evidence that hydrocarbons have passed through. **160** shows a limestone in which a few pores are filled with black hydrocarbon and others are lined by a thin coating of it. Examination of its relationship to the cement shows that the hydrocarbon entered the rock after an early generation of isopachous cement (marine?) and before the final filling of coarse blocky cement (meteoric).

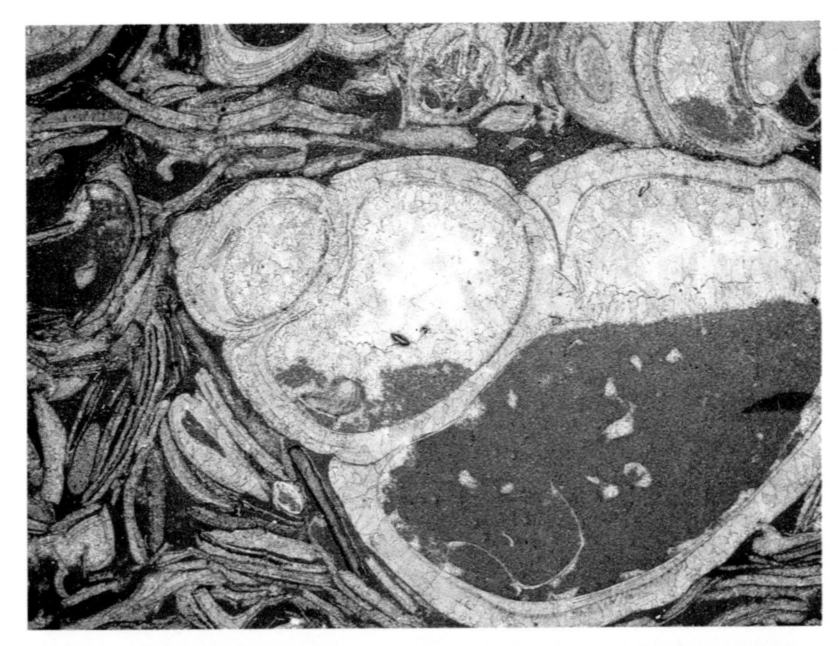

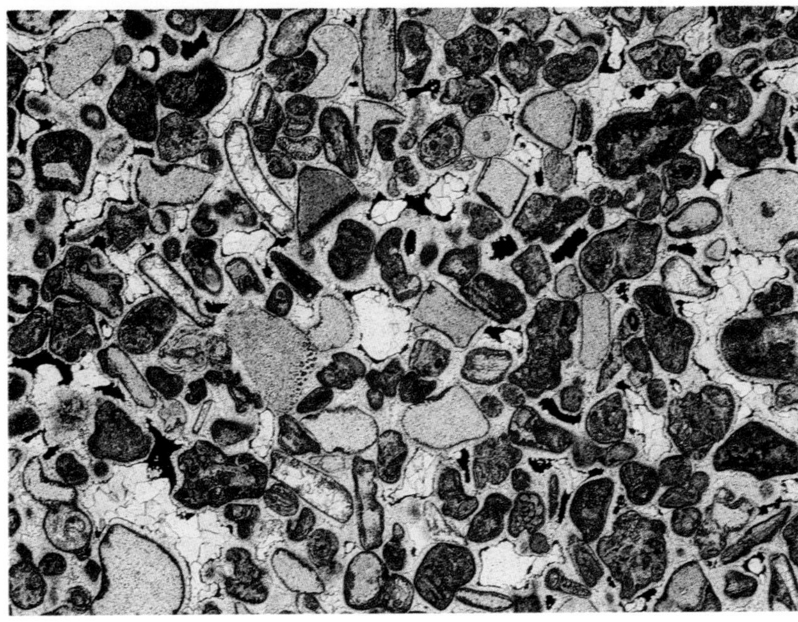

159*: Stained thin section, Purbeck Marble, Upper Jurassic, Dorset, England; magnification × 12, PPL.*
160*: Unstained thin section, Bee Low Limestone, Lower Carboniferous, Windy Knoll, Derbyshire, England; magnification × 16, PPL.*

Dolomitization

Introduction

Dolomite, $CaMg(CO_3)_2$, is a major component of limestones. It is usually secondary, replacing pre-existing carbonate minerals. Unlike calcite, it often occurs as euhedral rhomb-shaped crystals. However, since its optical properties are similar to those of calcite, it can be difficult to distinguish between the two. For this reason etching and staining of sections with Alizarin Red S is carried out (see p. 34).

Dolomitic rocks are classified according to their dolomite content as follows:

0 to 10% dolomite	limestone
10 to 50% dolomite	dolomitic limestone
50 to 90% dolomite	calcitic dolomite
90 to 100% dolomite	dolomite

Since the term dolomite is used for both the mineral and the rock, some workers prefer the term *dolostone* for the rock, although the term has not been universally accepted and is not employed here.

Dolomitization

(continued)

161 shows a dolomitic limestone containing 20 to 30% dolomite. The dolomite is unstained and occurs as euhedral rhomb-shaped crystals which contain inclusions, probably of calcite, and are thus cloudy. The unaltered limestone surrounding the dolomite is pink-stained, non-ferroan calcite and shows a patchy texture of micrite and sparite with few recognizable grains. This is a neomorphic fabric (p. 60).

162 shows a calcitic dolomite in which the original calcite matrix has been wholly replaced by dolomite (unstained) but the micritic allochems (peloids) have resisted dolomitization and are only partly replaced (dolomite unstained, calcite red). Where replacement is incomplete, euhedral rhomb-shaped crystals are visible. Where replacement is complete, crystals have grown together and the euhedral shape is lost.

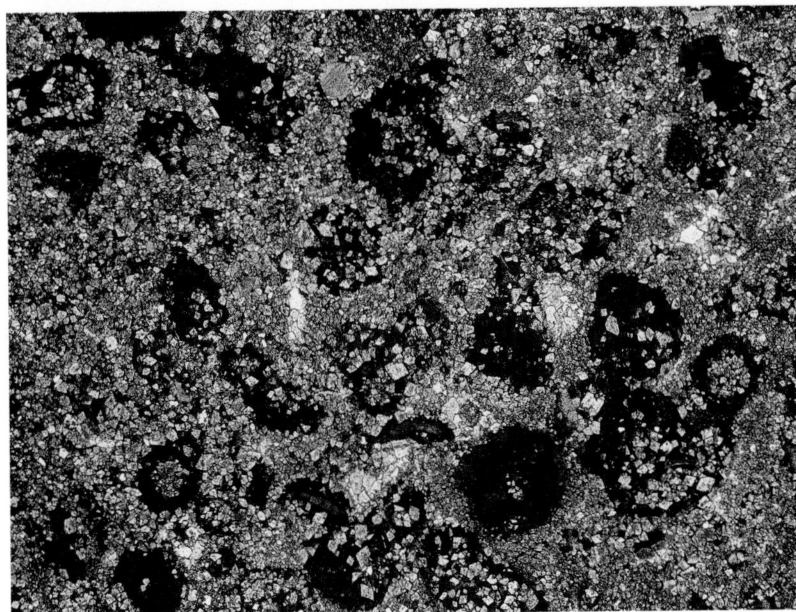

161: Stained thin section, Woo Dale Limestone, Lower Carboniferous, Derbyshire, England; magnification × 20, PPL.
162: Stained thin section, Middle Jurassic, Jebel Amsitten, Morocco; magnification × 14, PPL.

Dolomitization

(continued)

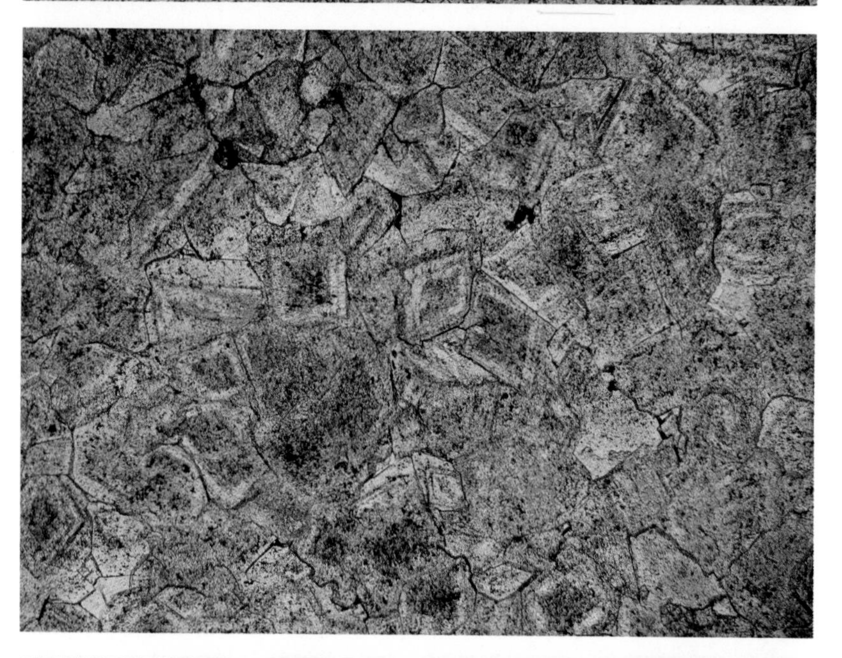

163 shows a sediment in which the original limestone has been totally replaced by dolomite. The result is a mosaic of anhedral crystals. Although the section was immersed in the staining solution, no stain at all is apparent, indicating the completeness of the replacement.

164 shows a dolomite in which the crystals are distinctly zoned. Although the crystal fabric is tightly interlocking, the rhombic shape of the dolomite crystals is clearly outlined by the zoning. The zoning may be partly caused by chemical differences in the dolomite but it is probably due mainly to varying amounts of foreign matter incorporated in the growing crystals.

The mineral dolomite may contain iron substituting for magnesium. When the iron content reaches 10 mole %, the term *ankerite* is used. **165** shows a ferroan dolomite approaching ankerite in composition. The ferroan nature of the mineral is shown by the turquoise stain colour (see p. 34), although the section was stained for a longer time than usual to enhance the colour, which is why the calcite present is red rather than pink-stained. The iron content of the dolomite is also shown by the dark brown margins to some crystals, where iron has been oxidized, producing limonite.

163: Stained thin section, Penmaen Burrows Limestone, Lower Carboniferous, Caswell Bay, South Wales; magnification × 43, PPL.
164: Unstained thin section, Woo Dale Dolomite, Lower Carboniferous, Woo Dale, Derbyshire, England; magnification × 56, PPL.
165: Stained thin section, Woo Dale Dolomite, Lower Carboniferous, Woo Dale, Derbyshire, England; magnification × 31, PPL.

Dolomitization

(continued)

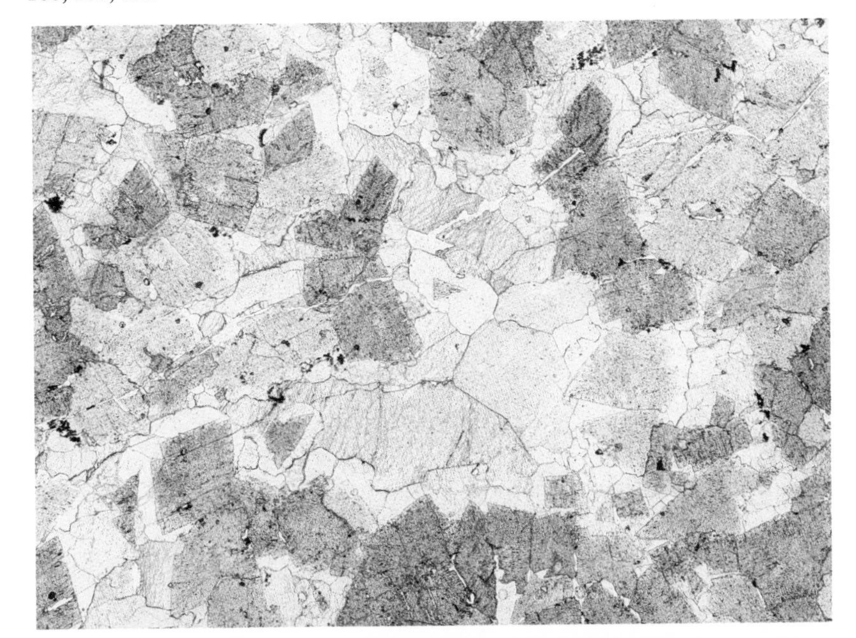

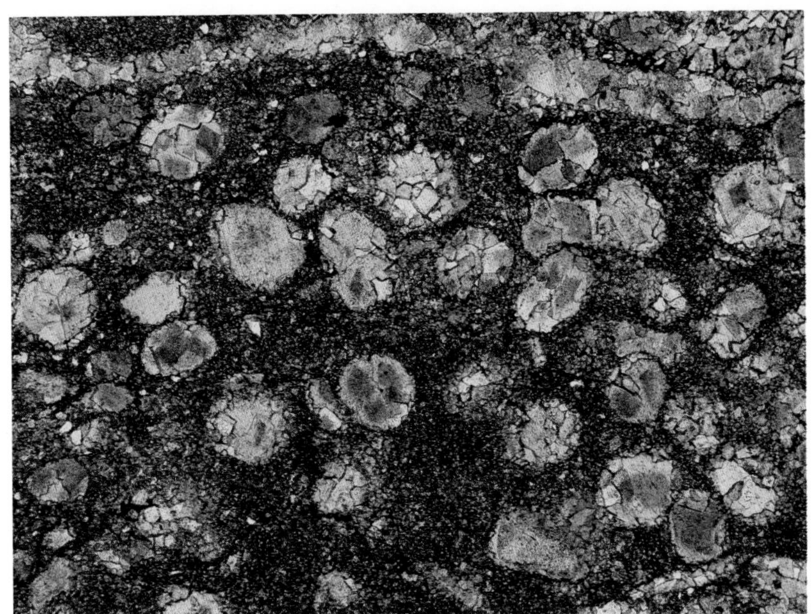

166 shows a loosely-interlocking network of euhedral dolomite crystals (unstained) with the intercrystal spaces infilled by a coarse sparry calcite cement (pink-stained).

Sometimes depositional textures are preserved in a rock despite complete replacement of the original sediment by dolomite. **167** shows a dolomite rock in which the matrix has been replaced by more finely crystalline dolomite than the allochems (perhaps originally ooids). The result is a 'ghost' texture.

168 shows a highly porous dolomite rock, some of the pores having been filled with a slightly ferroan calcite cement which is stained very pale mauve in thin section, but is too faint to reproduce well in the photograph. The dolomite is very fine-grained and the outlines of original allochems have been preserved as a ghost texture. Porosity in sediment replaced by dolomite is known as *intercrystal* porosity (Fig. F, see p. 65).

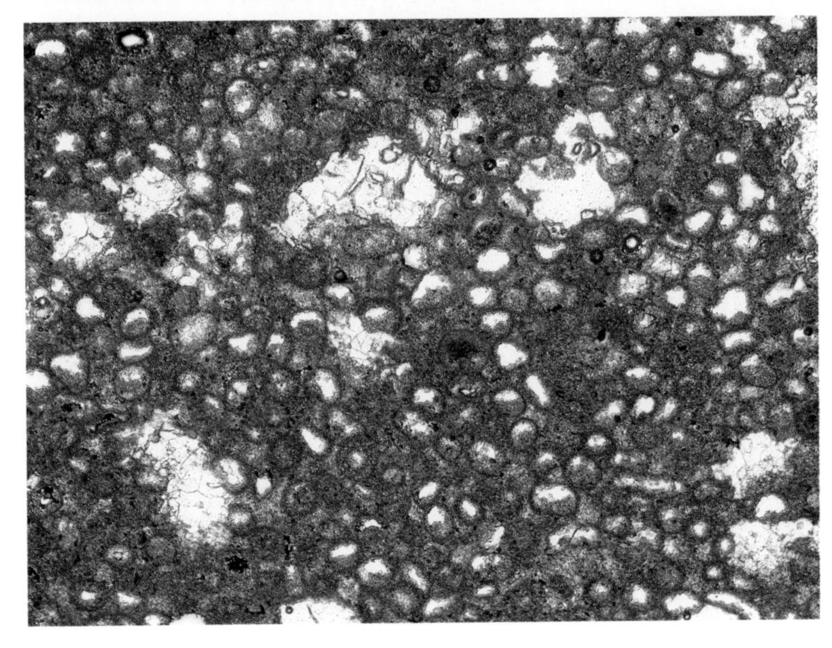

***166**: Stained thin section, Woo Dale Dolomite, Lower Carboniferous, Cunning Dale, Derbyshire, England; magnification × 15, PPL.*
***167**: Stained thin section, Ouanamane Formation, Middle Jurassic, Ouadirn, Western High Atlas, Morocco; magnification × 23, PPL.*
***168**: Stained thin section, Magnesian Limestone, Permian, South Yorkshire, England; magnification × 38, PPL.*

Dedolomitization

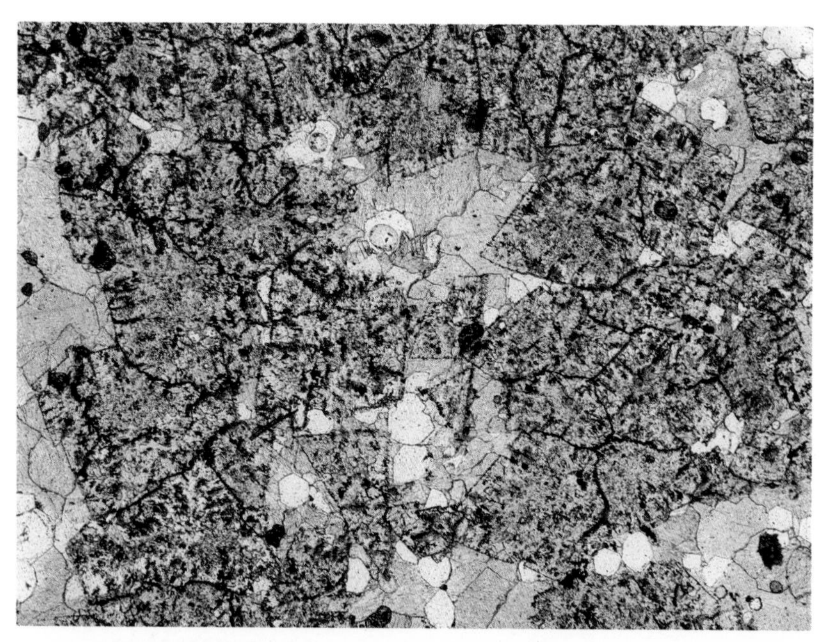

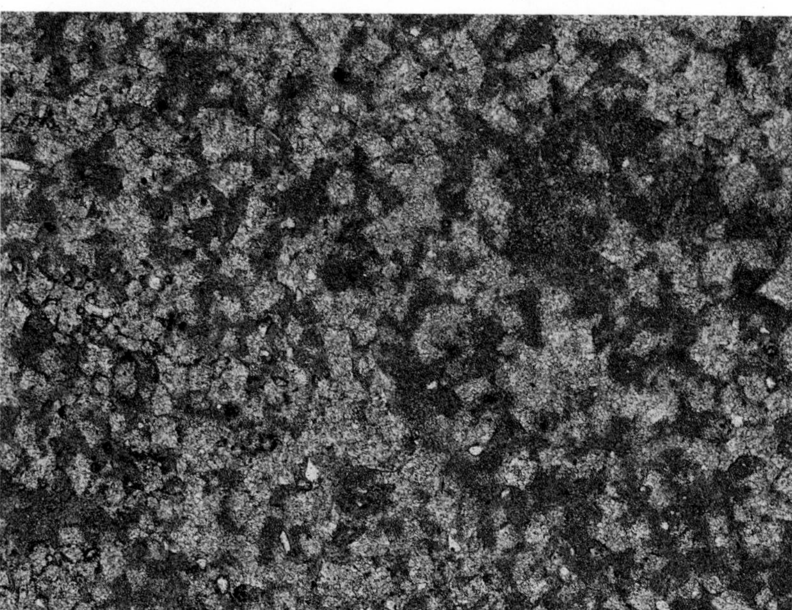

Dolomite may be replaced by calcite, usually by the action of oxidizing meteoric waters. This process of *dedolomitization* yields rhomb-shaped crystals of calcite or rhomb-shaped areas which comprise a mosaic of replacement calcite (dedolomite).

169 shows large rhomb-shaped areas which are now pink-stained calcite crystals. The morphology of these areas suggests that they were originally single dolomite crystals. Note that the 'dedolomite' is full of brown inclusions of iron oxides. This is a common feature since dedolomitization often occurs in oxidizing conditions, where any ferrous iron in the precursor dolomite is oxidized to produce iron oxides rather than be incorporated in the replacement calcite. A sparry calcite cement can be seen together with numerous hexagonal sections through an unstained mineral with low relief, which is authigenic quartz. Before dedolomitization this sediment would have been similar to the calcitic dolomite shown in **166**.

170 illustrates a dedolomite in which the former dolomite crystals have been replaced by a mosaic of small calcite crystals. Micritic calcite sediment occurs between the rhomb-shaped areas.

169: Stained thin section, Woo Dale Limestone, Lower Carboniferous, Cunning Dale, Derbyshire, England; magnification × 27, PPL.
170: Stained thin section, Upper Jurassic, Jebel Amsitten, Morocco; magnification × 42, PPL.
Dedolomite can also be seen in **137**.

Part 3

Other sedimentary rocks

Introduction

In this section we include photographs of thin sections of ironstones, evaporites, cherts, phosphorites and carbonaceous rocks. Even taken together, these rock types form a very small proportion of the total sedimentary record. However these groups of rocks have always attracted petrographic study out of proportion to their abundance, partly because they include rocks of great economic importance and partly because they show unusual features which have attracted considerable interest.

Ironstones

Sedimentary rocks with more than 15% iron, which have usually been worked as ores, are known as *ironstones*. Phanerozoic ironstones are usually local accumulations of fossiliferous oolitic deposits and are called *oolitic ironstones*, whereas Precambrian ironstones are much more extensive in area and comprise bedded alternations of iron minerals and silica. The latter are known as *banded iron formations*.

171 and **172** show a Jurassic oolitic ironstone. The olive-green mineral making up most of the ooids and the initial cementing material is *chamosite*. Chamosite is an iron silicate with a structure similar to that of chlorite. It occurs as fine-grained aggregates and, as the lower photograph shows, it has a birefringence of nearly zero. The brown-stained areas within the grains having high birefringence are *siderite*, iron carbonate, whereas the intergranular cement is calcite.

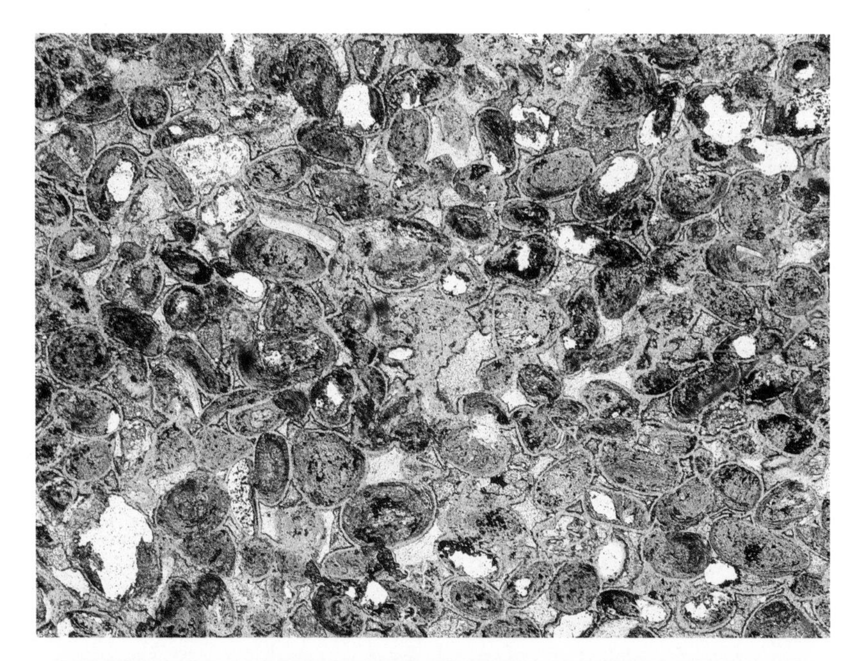

171 and 172: Northampton Sand Ironstone, Middle Jurassic, Northamptonshire, England; magnification × 37; 171 PPL, 172 XPL.

Ironstones

(continued)

Chamosite readily oxidizes to limonite, sometimes on the sea-floor soon after deposition, but more commonly, much later after uplift and weathering. **173** shows an ironstone containing both partially-altered chamosite ooids, which are yellow to golden-brown in colour with darker limonite zones, and totally-replaced grains of pure limonite which are almost opaque. The cement is calcite. The sediment is fairly soft and some grains have been plucked out during the making of the section, leaving holes which appear colourless in the photograph.

174 illustrates an ironstone in which ellipsoidal grains have been completely replaced by opaque limonite. The sample contains a few rounded shell fragments (e.g. left-hand side above centre) and scattered quartz grains, which although colourless and showing low relief, are sometimes surrounded by thin rims of dark limonite (examples can be seen near the centre). The cement is calcite.

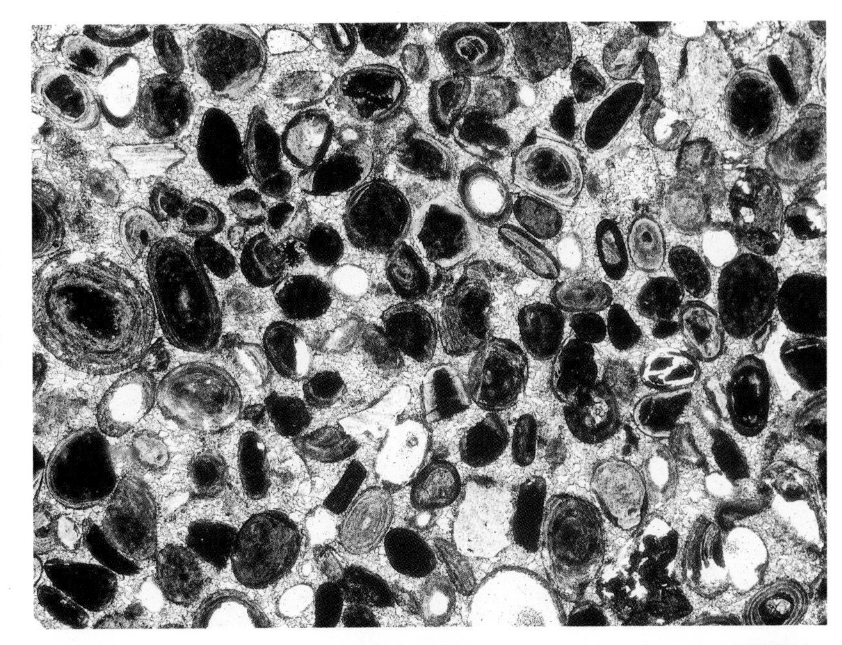

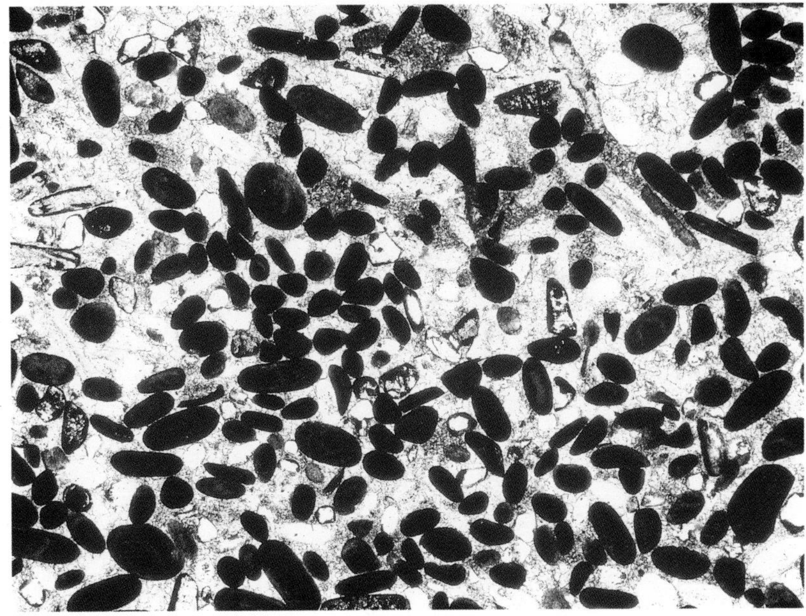

173: Northampton Sand Ironstone, Middle Jurassic, Northamptonshire, England; magnification × 37, PPL.
174: Frodingham Ironstone, Lower Jurassic, Scunthorpe, England; magnification × 35, PPL.

Ironstones

(continued)

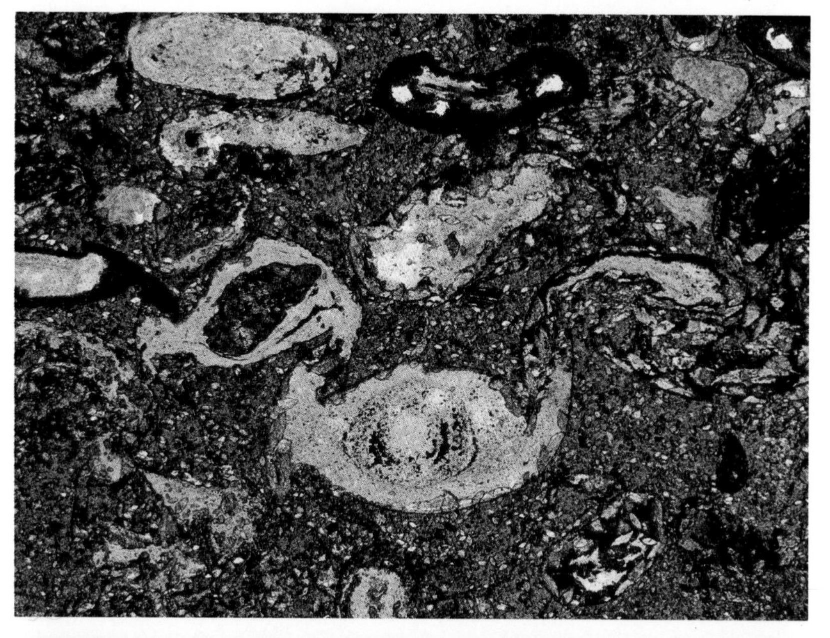

175 illustrates the abundant marine fauna found in many oolitic ironstones. The speckled fragments are echinoderm plates (a large example can be seen just below the centre). The concentric structure of the ooids is picked out by alternations of green chamosite and opaque iron oxides. The matrix contains iron oxides and small quartz grains (colourless, low relief).

Ooids in Phanerozoic ironstones are often squashed during compaction, suggesting that they were soft for some time after deposition. This contrasts with calcareous ooids which are rigid grains and thus retain their shape. **176** and **177** show chamosite ooids which have been squashed and hooked as a result of compaction. Such grains are known as *spastoliths*. The very low birefringence of the chamosite, with dark grey interference colours, is well seen in XPL. The matrix between the ooids contains dark, greenish-brown coloured chamosite mud and small roughly equidimensional crystals with high birefringence. In some places these can be seen to be rhombic in cross-section and are *siderite* (iron carbonate). Larger brown-stained siderite crystals can be seen replacing the margins of some of the ooids.

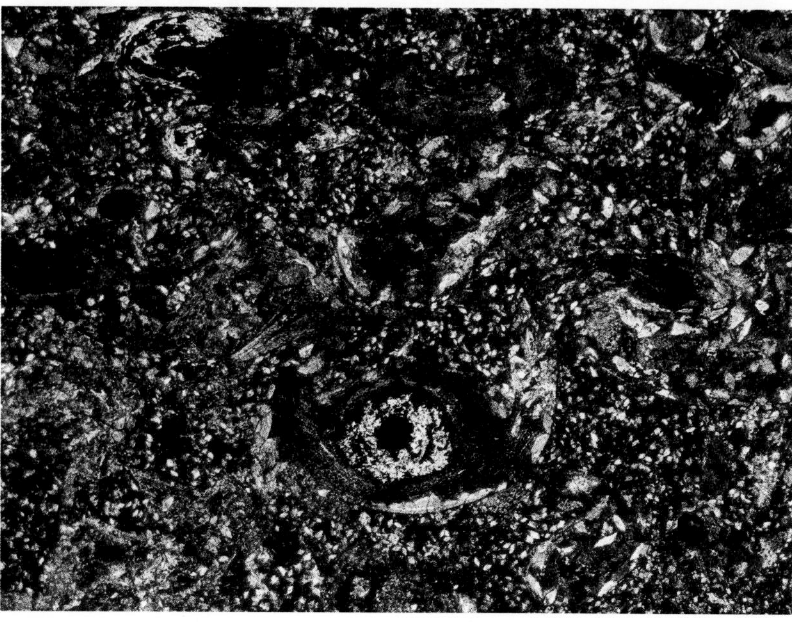

175: Lower Jurassic, Skye, Scotland; magnification × 16, PPL.

176 and 177: Raasay Ironstone, Lower Jurassic, Raasay, Scotland; magnification × 43; ***176*** *PPL,* ***177*** *XPL.*

Ironstones

(continued)

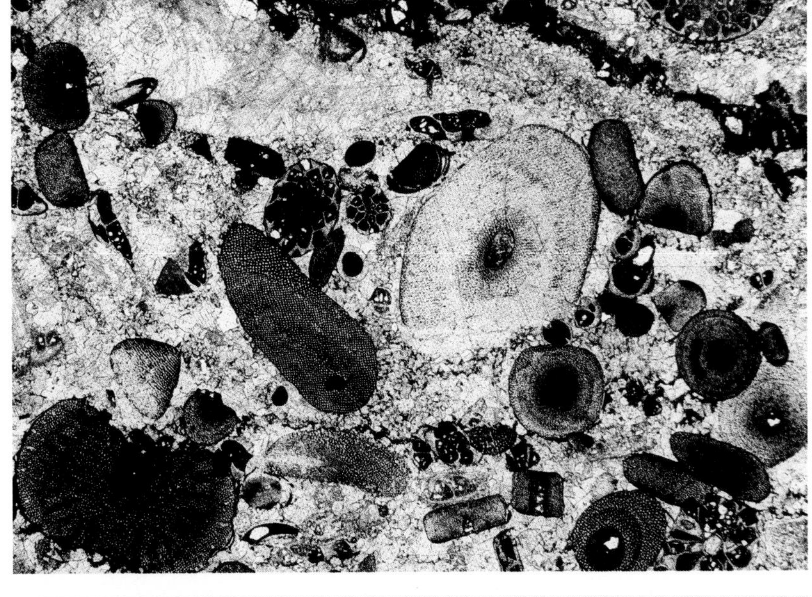

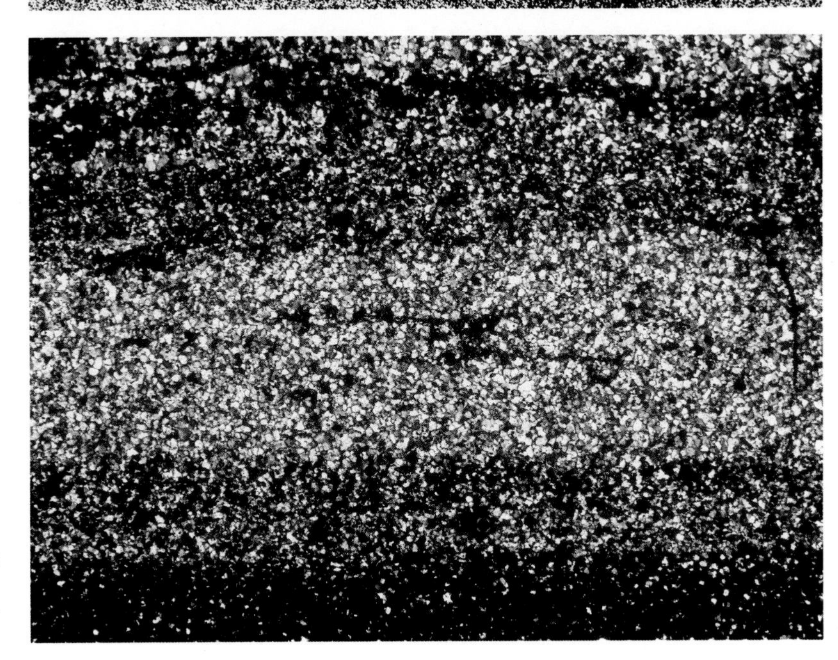

178 shows a limestone in which bioclasts have become impregnated by opaque iron oxide. The iron oxide has infilled the pores in crinoid fragments and partly replaced their skeletons, leading to the development of a distinctive reticulate structure. Bryozoans also have been impregnated by iron oxide and two examples can be seen just above and left of the centre of the photograph.

179 shows a thin section of a Precambrian banded ironstone comprising alternations of dark, iron oxide-rich layers and colourless chert layers. **180** is a higher magnification view of part of the same section taken with crossed polars and showing the fine-grained quartz which makes up the chert layers.

178: Rhiwbina Iron Ore, Lower Carboniferous, South Wales; magnification × 20, PPL.
179 and 180: Precambrian, Transvaal; 179, magnification × 9, PPL; 180, magnification × 32, XPL.

Cherts

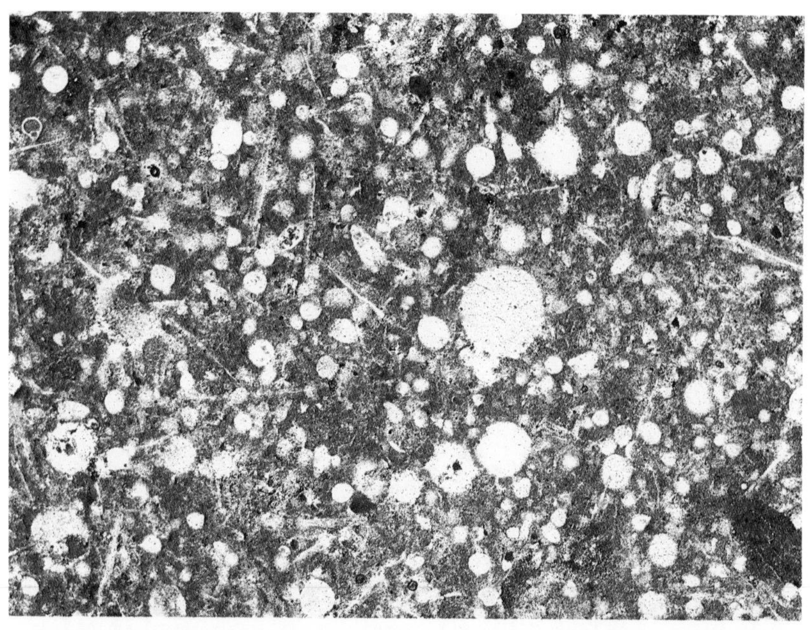

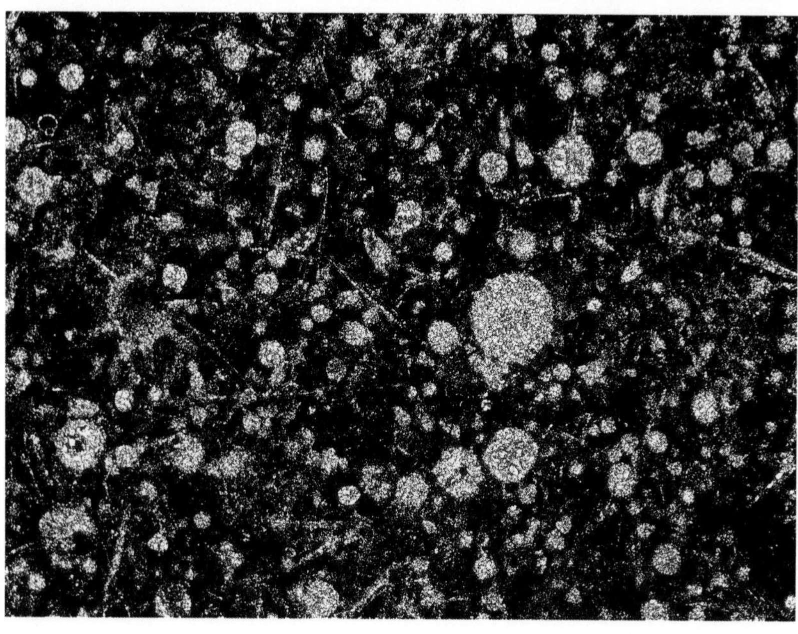

Cherts are rocks composed of authigenic silica usually in the form of fine-grained quartz. Cherts may be primary, in which case most of the silica is in the form of hard parts of siliceous organisms such as radiolaria, diatoms and some sponges. Much chert, however, is secondary, usually replacing limestone. Scattered grains of authigenic replacement quartz, often hexagonal in cross-section, are not uncommon in limestones and are illustrated in **73** and **169**.

Radiolaria are siliceous microfossils which accumulate in sediments of the deep ocean floor. **181** and **182** show a radiolarian chert with spherical radiolaria skeletons and their long thin spines. The matrix contains fine-grained iron oxide, hence the red-brown colour. Radiolaria are originally opaline silica, an isotropic form containing water. This has been converted to fine-quartz (microquartz) showing low first-order interference colours in XPL.

*181 and 182: Lower Cretaceous, Greece; magnification × 32; **181** PPL, **182** XPL.*
*Calcite casts of radiolaria are shown in **89**.*

Cherts

(continued)

183 and **184** illustrate features found in limestones, ironstones and terrigenous rocks as well as in cherts. The sediment contains terrigenous quartz grains, both monocrystalline and polycrystalline (p. 5), which are clear in PPL. It contains bioclasts, including an endopunctate brachiopod shell (upper left) and echinoderm plates impregnated with iron oxide (e.g. upper right). These are composed of calcite and have been stained pale pink by the use of Alizarin Red S (see p. 34). The sediment also contains silicified grains which are unstained and brownish in colour. These include both structureless examples, which are seen to be made up of microquartz in varying orientations (e.g. below the centre), and those with a nucleus of terrigenous quartz and an outer zone of silica with an oolitic structure (e.g. left of centre, near the top). These are interpreted as silicified ooids. Evidence from the XPL photograph suggests that in some of these at least, the silica replacing the ooid cortex has grown syntaxially with the terrigenous quartz nucleus (shown by uniform interference colour) e.g. the grain just below half-way up, right of centre.

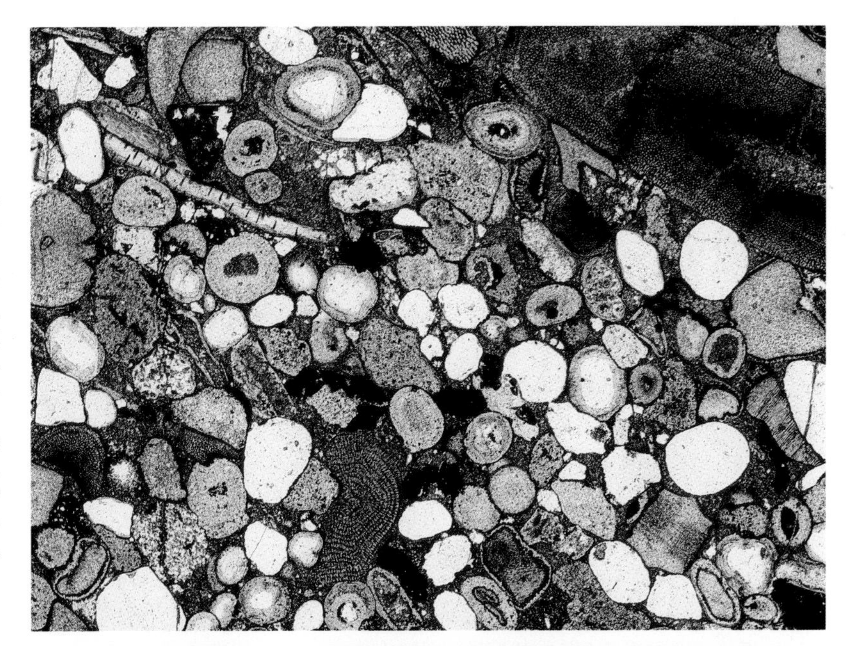

***183** and **184**: Stained thin section, Carboniferous, Middle Atlas, Central Morocco; magnification × 17; **183** PPL, **184** XPL.*

Cherts

(continued)

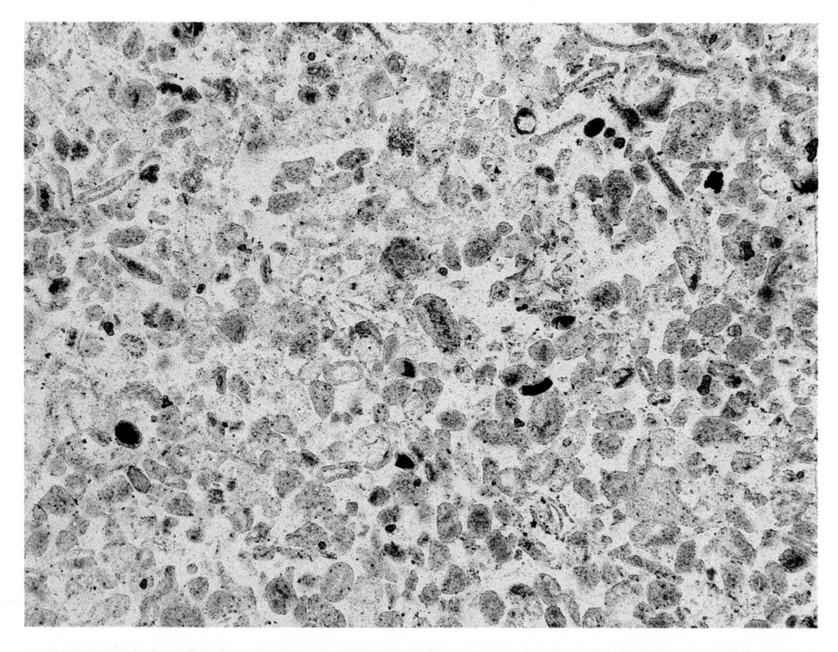

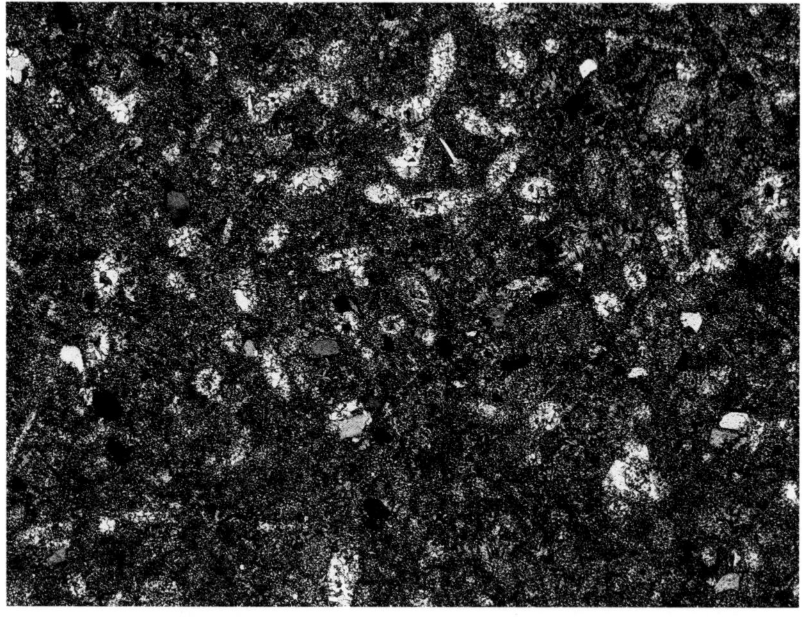

185 and **186** show a silicified limestone in which silicification is not quite complete. The brownish grains seen in PPL are unaltered calcite, as indicated by their high birefringence when viewed with crossed polars. Those grains which are clear in PPL show first-order interference colours when the polars are crossed, and have been entirely replaced by quartz. Although the limestone has undergone considerable alteration, the ghost texture visible shows that the original sediment comprised small rounded grains (peloids) and a few shell fragments.

185 and 186: Age and locality unknown; magnification × 22; 185 PPL, 186 XPL.

Cherts

(continued)

187 and **188** show a variety of quartz types. The circular to elliptical areas of fine quartz (microquartz) may be the original sediment grains replaced by silica. The surrounding areas consist of clear and cloudy zones of radial-fibrous quartz, known as *chalcedonic quartz*. The final generation of brown silica illustrates most clearly the radial-fibrous structure. Chalcedonic quartz is often a pore-fill rather than a replacement. This is supported from the evidence of the sample illustrated, in that there are straight boundaries between adjacent growth of chalcedonic quartz and triple points where three growths meet. These polygonal boundaries are characteristic of radial-fibrous, pore-filling cements. The upper right of the photographs shows coarse equant quartz (macroquartz) which contains inclusions of highly birefringent carbonate, indicating that the silica has probably replaced limestone.

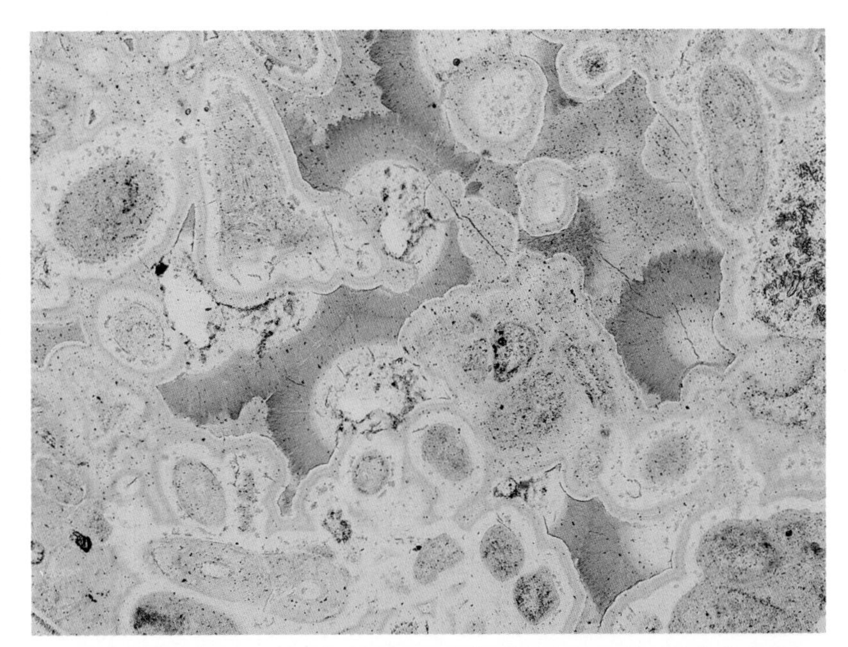

***187** and **188**: Upper Jurassic, Dorset, England; magnification × 43; **187** PPL, **188** XPL.*

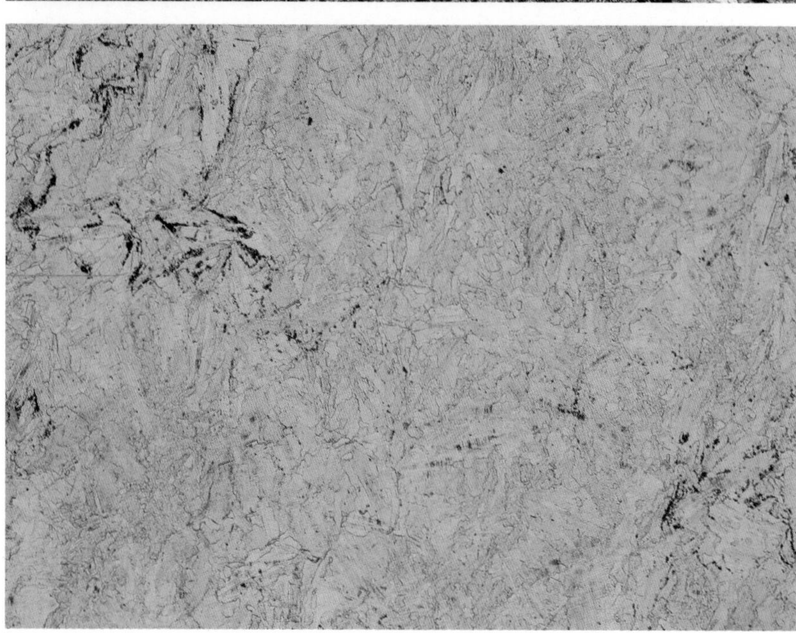

Evaporites

Evaporites are rocks composed of minerals which precipitate from natural waters concentrated by evaporation. Although only a few minerals are abundant in marine evaporite deposits, complex textures may develop as a result of the replacement of one mineral by another during diagenesis. On evaporation of seawater, the first minerals to precipitate after carbonate are the calcium sulphates. The hydrous form, *gypsum*, ($CaSO_4.2H_2O$) occurs only near the earth's surface, whereas *anhydrite* ($CaSO_4$) is formed at the surface and also replaces gypsum at depth.

189 and **190** show laths of gypsum partially filling a cavity in a dolomite rock. The dolomite shows the very high relief and strong birefringence of a carbonate, whereas the gypsum shows low relief and weak birefringence. The photograph taken with crossed polars shows typical gypsum interference colours, up to first-order pale grey.

191 and **192** show a thin section through a sediment composed almost entirely of anhydrite. It can be distinguished from gypsum by its higher relief and stronger birefringence. In the example shown, the anhydrite is in the form of laths with a radiating habit. The view taken with crossed polars shows the bright second-order interference colours characteristic of anhydrite.

193 and **194** show a sediment composed of gypsum and dolomite. The dolomite is very fine-grained and almost opaque in the photograph. The gypsum is in two forms. At the base and top of the photograph it is in the form of a network of irregular crystals, whereas in the centre it is in the form of fibres aligned at right angles to the bedding. The former type is characteristic of gypsum replacing anhydrite, whereas the fibrous gypsum is filling a vein running parallel to bedding.

Evaporites

(continued)

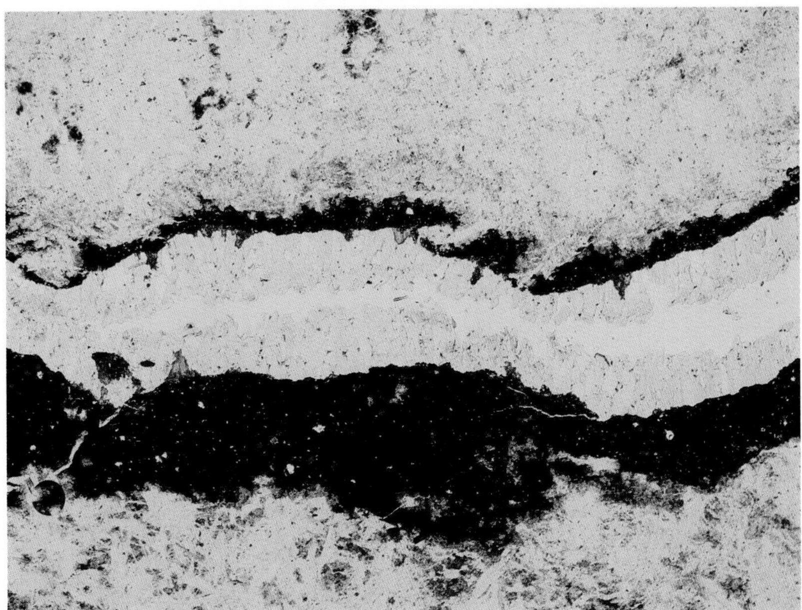

***189** and **190**: Carboniferous Limestone, Taff's Well, South Wales; magnification × 20; **189** XPL, **190** XPL.*
***191** and **192**: Permian, Billingham, Teesside, England; magnification × 16;* **191** *PPL,* ***192** XPL.*
***193** and **194**: Permian, Billingham, Teesside, England; magnification × 9; **193** PPL, **194** XPL.*

Evaporites

(continued)

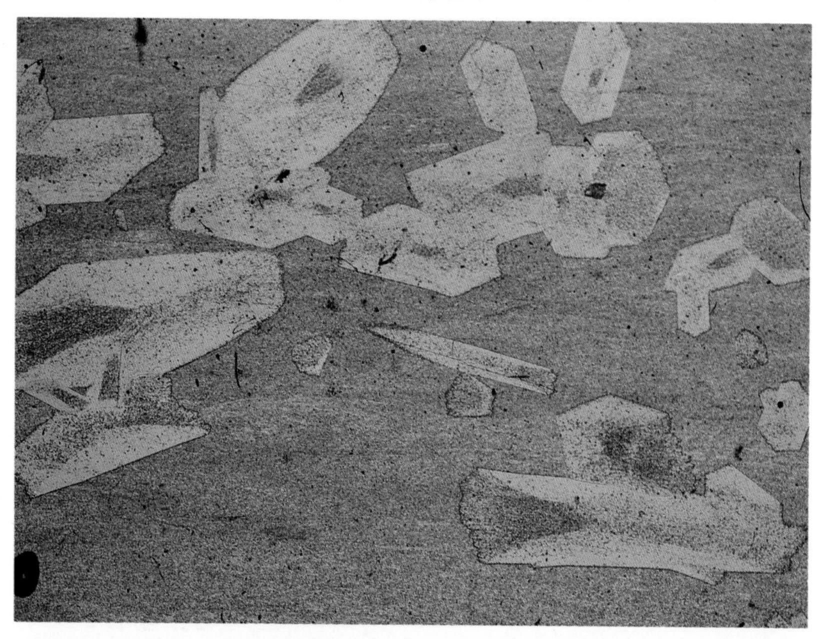

Gypsum may replace anhydrite on uplift of evaporite sequences and when removal of the overburden brings them near the surface. Textures are often of the type shown in **193** and **194**, with small irregular gypsum crystals, but sometimes large euhedral crystals form. **195** and **196** show gypsum porphyroblasts replacing fine-grained (aphanitic) anhydrite. Note the six-sided gypsum crystals with low relief and first-order interference colours, contrasting with the anhydrite showing moderate relief and bright second-order interference colours. Note that the distribution of relict anhydrite inclusions within the gypsum porphyroblasts has sometimes led to the development of a texture similar to 'hourglass' zoning, a feature found in some minerals in igneous and metamorphic rocks.

***195** and **196**: Permian, South Durham, England; magnification × 8; **195** PPL, **196** XPL.*

Evaporites

(continued)

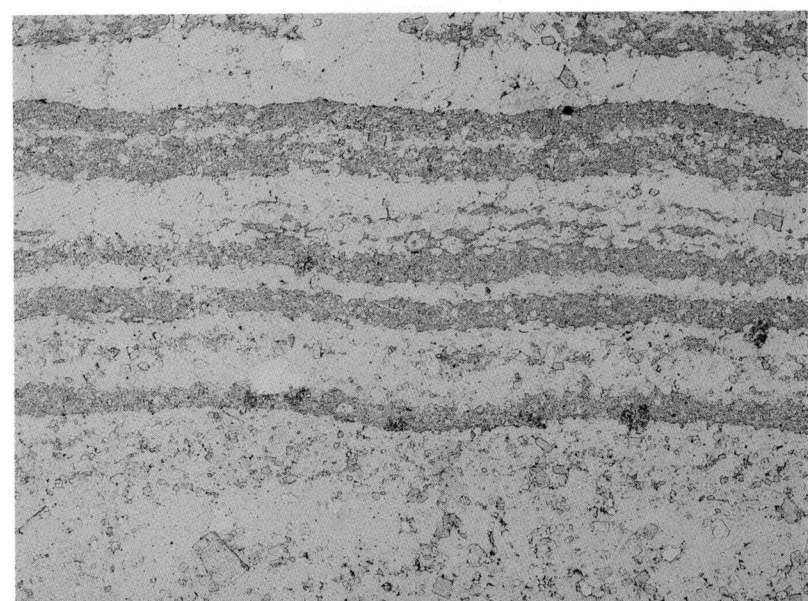

The two most common chloride minerals in evaporite sequences are *halite* (NaCl) and *sylvite* (KCl). **197** shows these minerals together. The refractive index of halite is close to that of the mounting medium, so that it shows very low relief, whereas the sylvite has a moderate negative relief. Some of the sylvite is reddish-brown in colour owing to the presence of a small amount of hematite, although the crystal in the lower right-hand corner is hematite-free. The perfect {100} cleavage of both minerals is visible in some crystals and the halite shows some evidence of zoning. Both halite and sylvite are cubic and thus isotropic.

198 and **199** show a layered anhydrite-halite rock. The thin layers of fine-grained anhydrite show moderate relief in PPL and bright second-order interference colours with polars crossed. The halite has low relief and is isotropic. The halite layers also contain scattered rectangular anhydrite crystals.

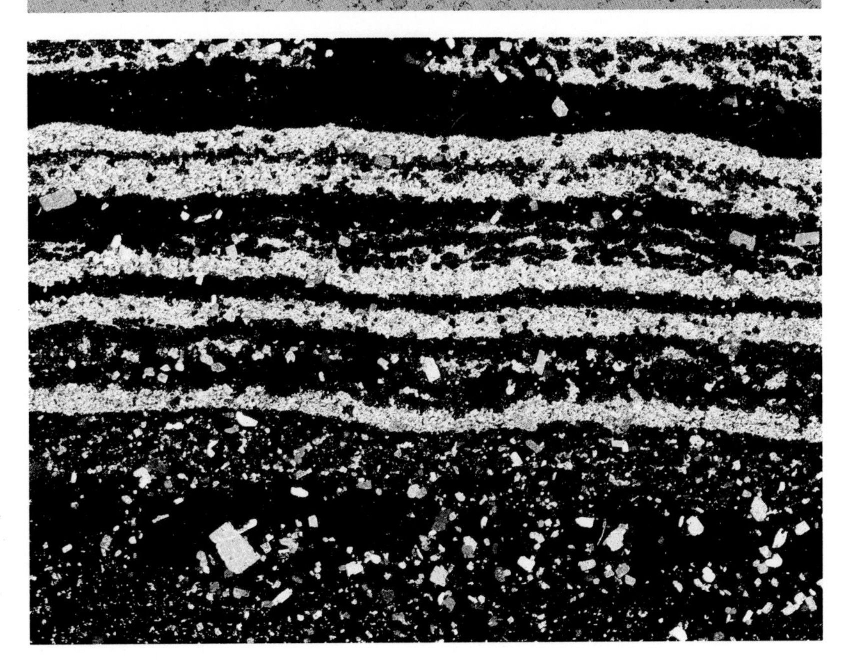

197: *Permian, Fison's Borehole, Robin Hood's Bay, North Yorkshire, England; magnification × 20, PPL.*
198 *and* ***199***: *Permian, Fordon No. 1 Borehole, Scarborough, North Yorkshire, England; magnification × 9;* ***198*** *PPL,* ***199*** *XPL.*

Evaporites

(continued)

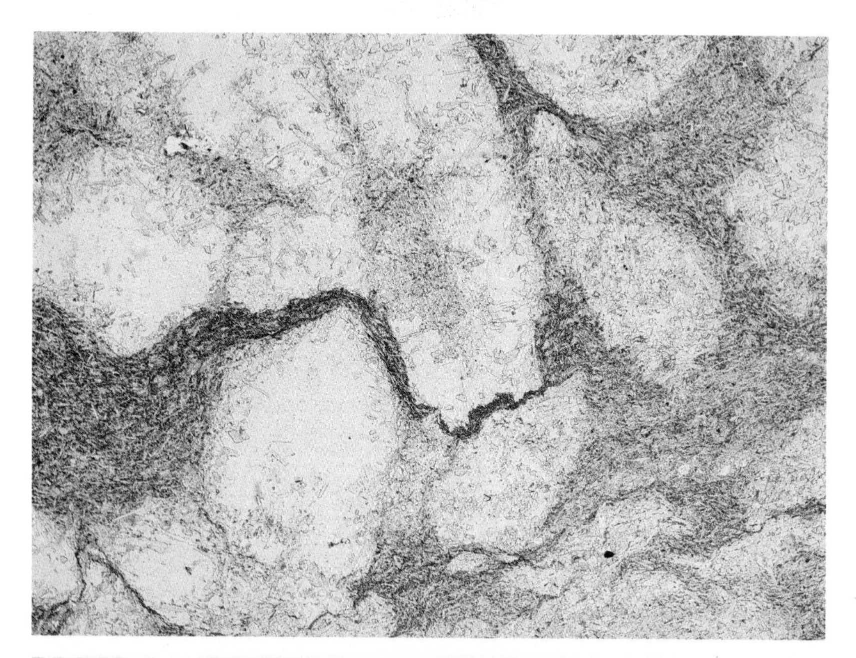

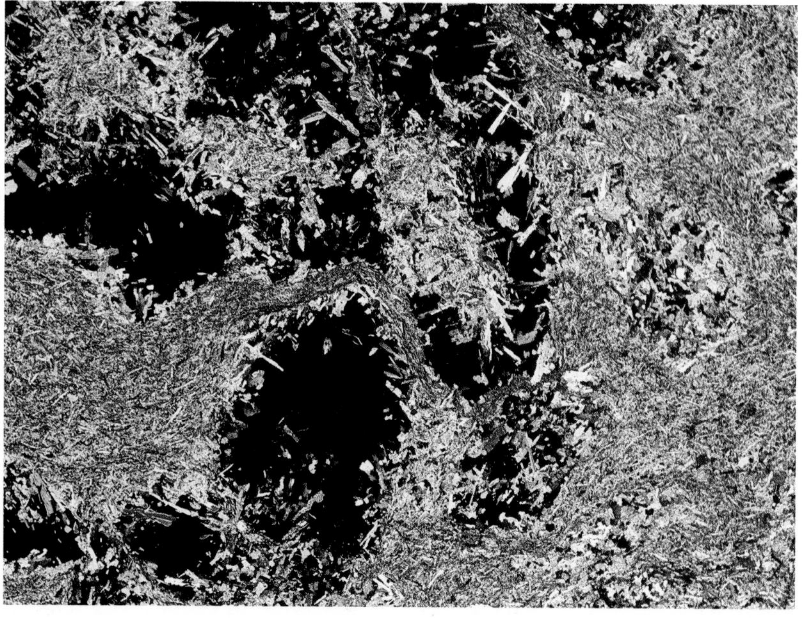

200 and **201** show an evaporite in which the minerals now present are halite (low relief, isotropic) and anhydrite (moderate to high relief, second-order interference colours). There is also some carbonate between the small anhydrite crystals which is too fine-grained to be resolved at the magnification shown here. The irregular six-sided shapes which are now composed principally of halite and scattered anhydrite laths, have the form of gypsum porphyroblasts (see **195** and **196**). They are thus interpreted as gypsum crystals which have been replaced. The gypsum itself was probably replacing anhydrite; this illustrates the complexity of diagenetic reactions which may occur in evaporites.

***200** and **201**: Permian, Hawsker Borehole, near Whitby, North Yorkshire, England; magnification ×8; **200** PPL, **201** XPL.*

Evaporites

(continued)

Polyhalite, $K_2MgCa_2(SO_4)_4.2H_2O$, is a common mineral in some marine evaporite sequences. **202** shows a rock composed essentially of polyhalite and halite. Both minerals have a similar relief and crystals are not easily distinguishable in PPL. Thus only a view taken with crossed polars is shown. The halite is isotropic and so appears black. The polyhalite is partly fine-grained and partly coarse-grained. The larger crystals show simple twinning. Polyhalite has a fairly low birefringence, and interference colours up to low second-order can be seen.

203 and **204** show a rock which is predominantly a fine-grained polyhalite. Two large porphyroblasts of anhydrite with a typical lath shape can also be seen. The distinct difference in relief of the two anhydrite crystals results from their different orientations with respect to the polarizer. The anhydrite crystal near the lower edge shows the rectangular cleavage. The anhydrite is being replaced by the polyhalite. The dark spots seen in **203** are granules of bituminous carbonate.

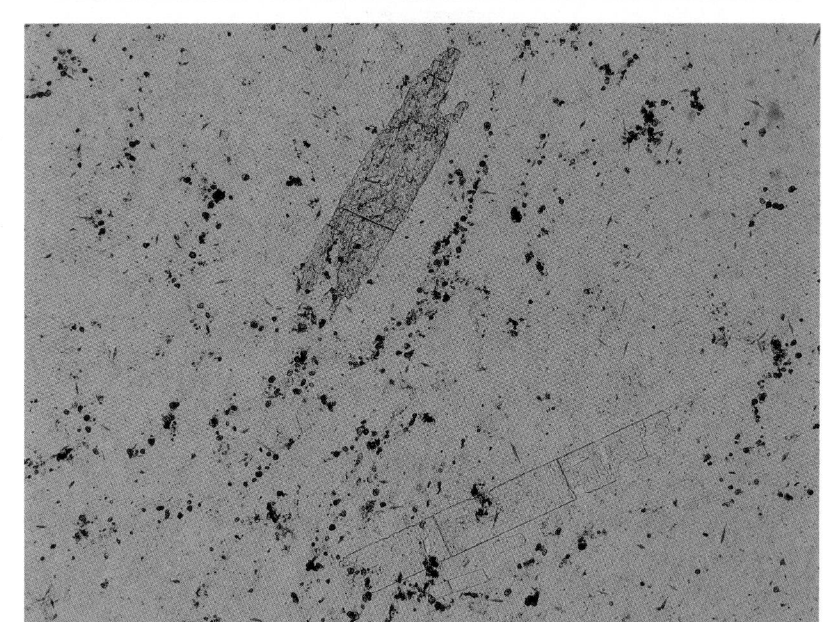

***202**: Permian, Fordon No. 1 Borehole, Scarborough, North Yorkshire, England; magnification × 16, XPL.*
***203** and **204**: Permian, Aislaby Borehole, near Whitby, North Yorkshire, England; magnification × 20; **203** PPL, **204** XPL.*

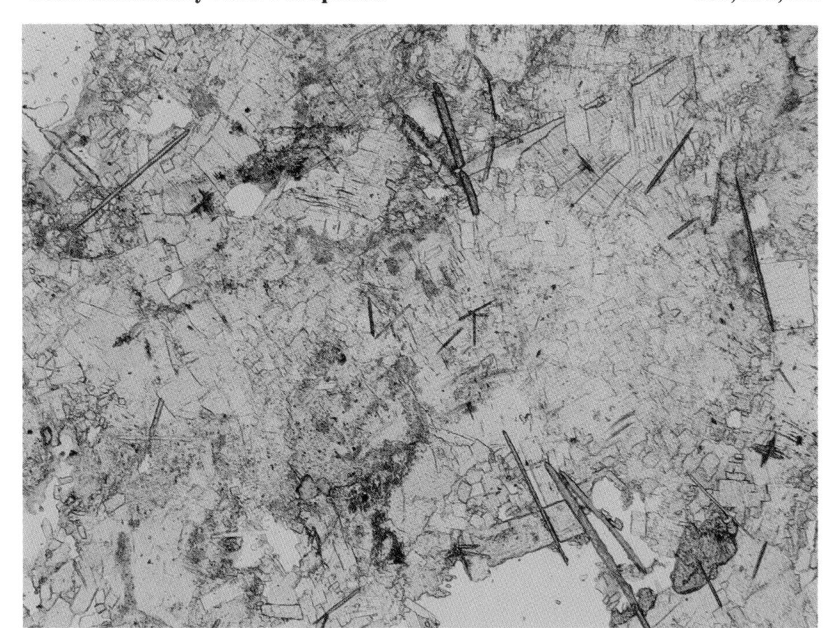

Evaporites

(continued)

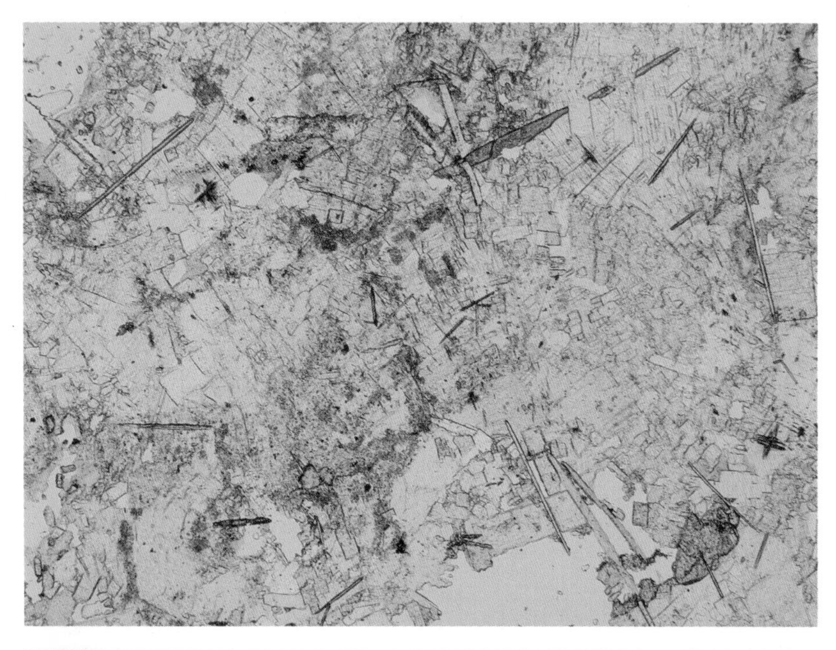

205, **206** and **207** show a rock composed of anhydrite and long narrow crystals of a carbonate mineral. The anhydrite occurs as separate rectangular crystals and as sheaves of sub-parallel laths. One of these sheaves, having first-order grey interference colours, can be seen towards the lower, right-hand corner of the photograph. Most of the anhydrite crystals show two cleavages at 90°, and bright second- and third-order interference colours. The two photographs taken in PPL show the marked change in relief of the carbonate mineral, achieved by rotating the polarizer through 90°. **207**, taken under crossed polars, shows that this mineral has very high order interference colours. It has been identified as *magnesite*, $MgCO_3$.

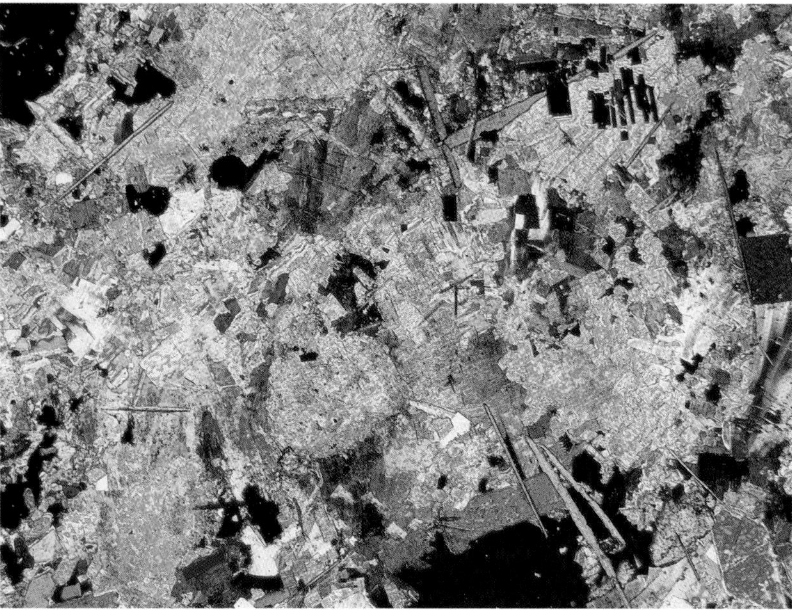

***205**, **206** and **207**: Permian, Aislaby Borehole, near Whitby, North Yorkshire, England; magnification × 20; **205** and **206** PPL, **207** XPL.*

Evaporites

(continued)

Carnallite ($KCl.MgCl_2.6H_2O$) is one of the most soluble of evaporite minerals and thus preparation of thin sections containing it is difficult. **208** and **209** show a thin section of an evaporite rock which is slightly thicker than the usual 30 μm. The carnallite is showing bright interference colours. Crystals in the centre and in the lower part of the field of view show the multiple twinning which is a characteristic of this mineral. The isotropic mineral is halite and the sediment also contains small rectangular crystals of anhydrite showing high birefringence.

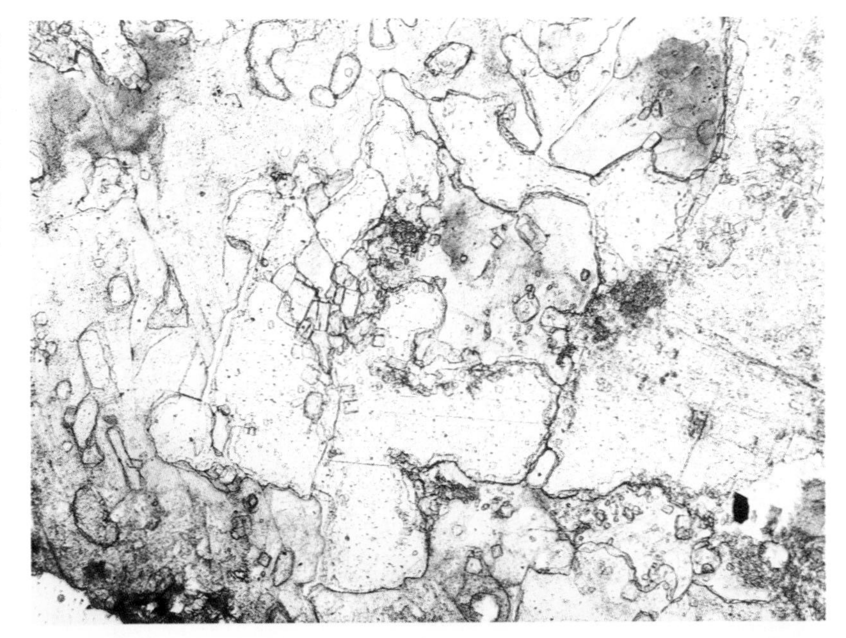

***208** and **209**: Permian, Fison's Borehole, Robin Hood's Bay, near Whitby, North Yorkshire; magnification × 19; **208** PPL, **209** XPL.*

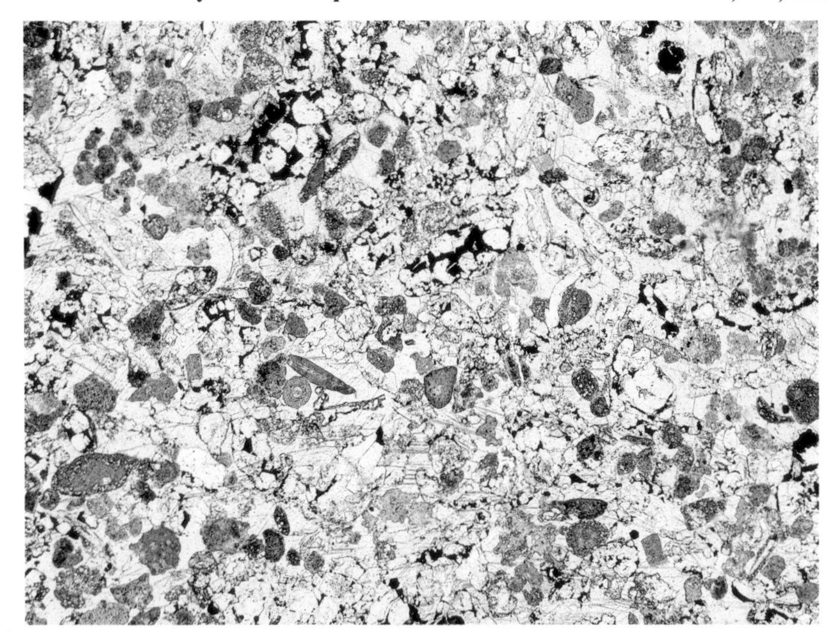

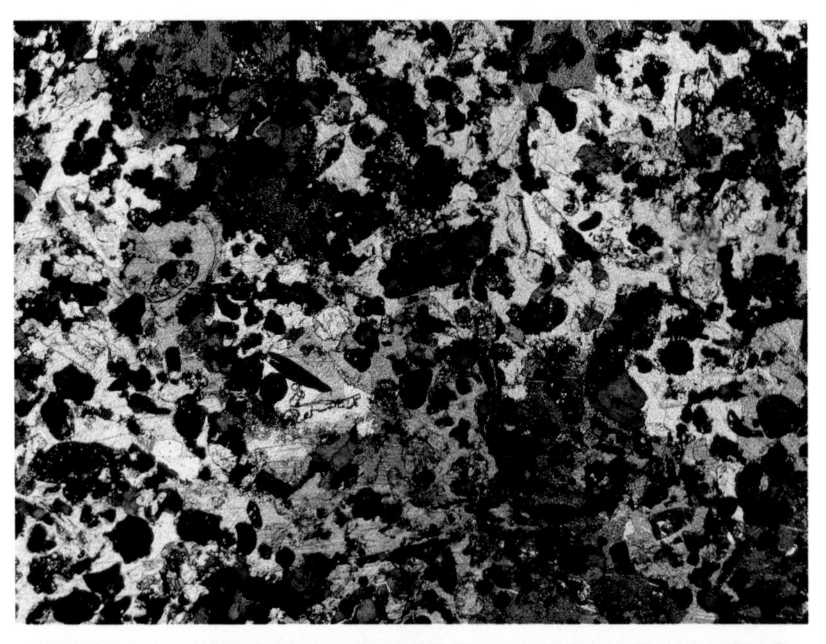

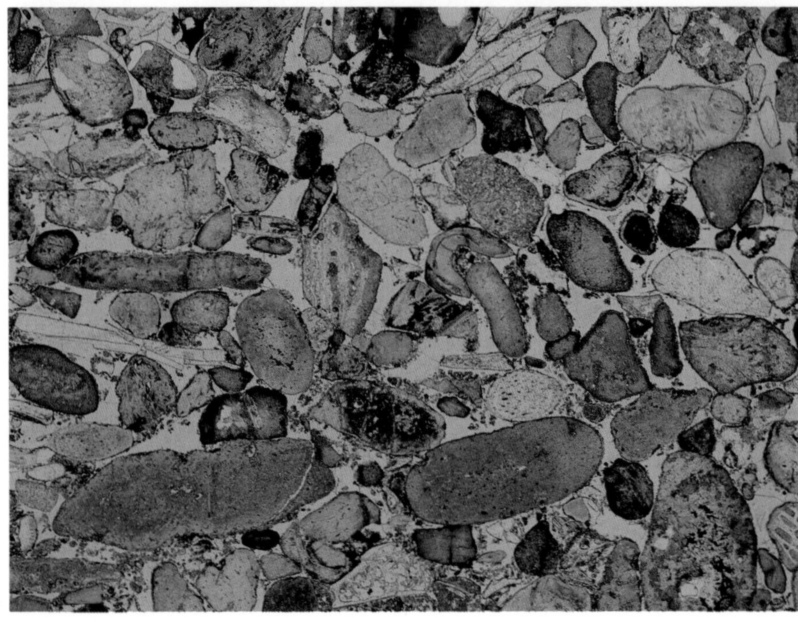

Phosphatic sediments

Some marine sediments contain authigenic phosphate, usually in the form of a cryptocrystalline carbonate hydroxyl fluorapatite, known as *collophane*. It commonly occurs as ooids and pellets, or as biogenic material such as fish teeth and scales or bone fragments. Sedimentary rocks rich in phosphate are called *phosphorites*.

210 and **211** show a phosphorite containing small greyish-brown pellets of collophane set in a coarse calcite cement. The view taken with polars crossed shows the isotropic nature of collophane and the high-order interference colours of the calcite. Visible in the lower left-hand quadrant is a grain of secondary quartz showing first-order grey interference colours.

212 and **213** illustrate a phosphorite in which the grains are principally brown-coloured pellets of isotropic collophane. The colourless fragments, some of which show very weak birefringence, are also phosphate. Some show a trace of internal structure (e.g. the grain to the right and below the centre). These are probably fish teeth and bone fragments. In contrast to the phosphorite shown in **210** and **211** where the cement is calcite, the cement here is fine-grained quartz.

214 and **215** show a limestone which has been stained with Alizarin Red S and potassium ferricyanide (see p. 34). The fragments are mainly oysters (pink-stained) and sections of hollow calcareous worm tubes (mauve) set in a ferroan calcite cement (blue). The sediment contains rounded pebbles of brown-coloured collophane (isotropic), containing small quartz grains and scattered pellets of green-coloured glauconite (see p. 17). The sediment also contains a few large, rounded quartz grains (e.g. the grain in the upper right corner).

Phosphatic sediments

(continued)

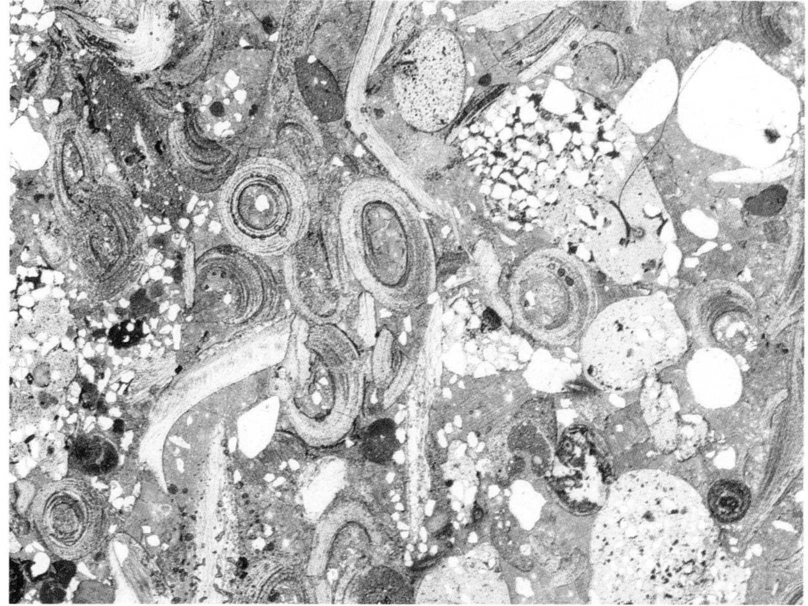

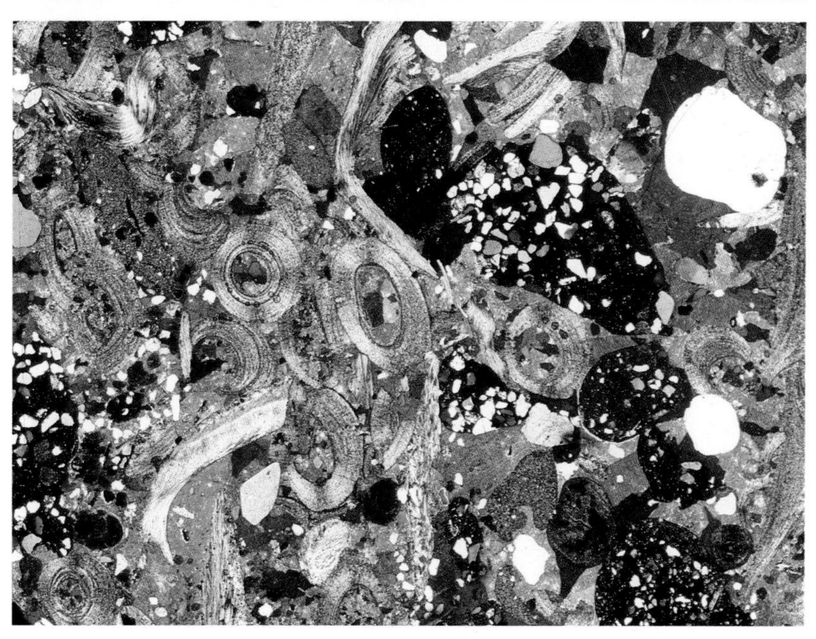

***210** and **211**: Carboniferous, Roadford, Co. Clare, Republic of Ireland; magnification × 23; **210** PPL, **211** XPL.*
***212** and **213**: Duwi Phosphate Formation, Eocene, Red Sea Coast, Egypt; magnification × 10; **212** PPL. **213** XPL.*
***214** and **215**: Tour de Croi Nodule Bed, Upper Jurassic, Wimereux, France; magnification × 11; **214** PPL, **215** XPL.*

Coals and coal balls

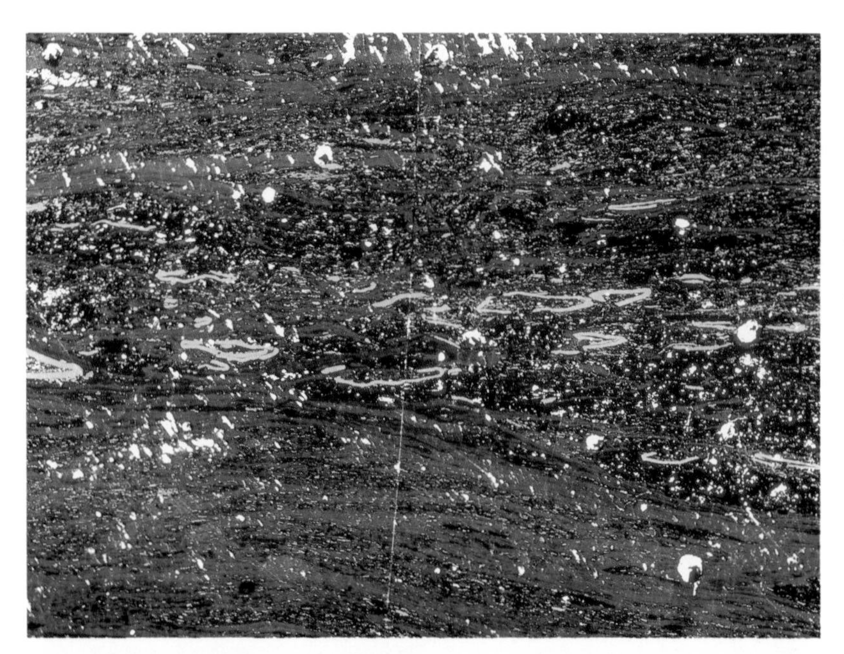

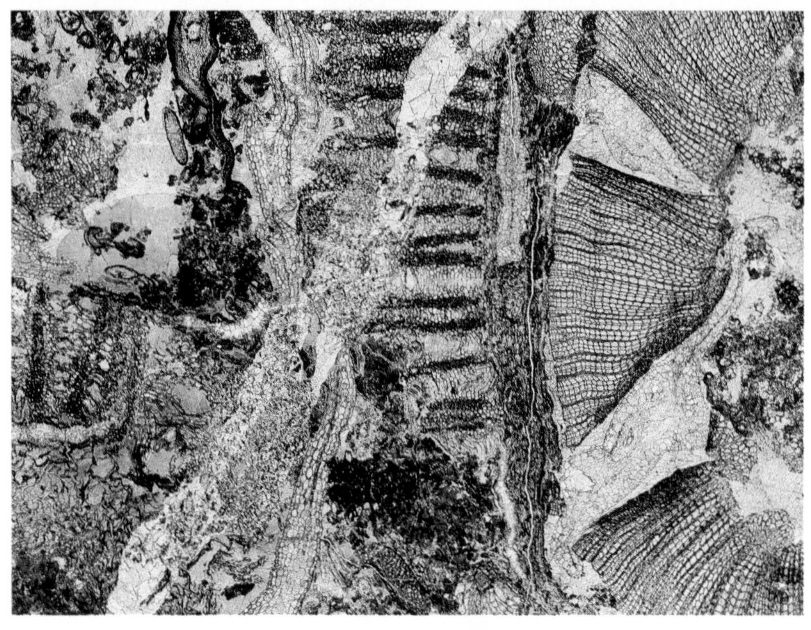

Coals are usually examined microscopically under high power in reflected light using oil-immersion objectives and therefore their detailed petrology is beyond the scope of this book. **216** is a thin section of a coal viewed in transmitted light. The photograph shows *durain*, the dull material in coal, composed of the more resistant plant matter. The bright yellow fragments are spore-cases distorted during compaction of the coal.

Coal balls are carbonate concretions formed before compaction of the plant material in coal. They may be up to a few tens of centimetres across and **217** shows a thin section of part of one. The photograph shows that they contain well-preserved cellular tissue of plants.

216: Coal Measures, Upper Carboniferous, England; magnification × 14, PPL.
217: Coal Measures, Upper Carboniferous, Lancashire, England; magnification × 9, PPL.

Appendix 1

Preparation of a thin section of a rock

It is sometimes believed that complex and expensive equipment is required for making thin sections of rock of standard thickness of 0.03 mm, but as the following instructions indicate, this is not the case. Thin sections can be made by the amateur with a little patience and perseverance. If a diamond saw is available to cut a slab of rock 1–2 mm in thickness, the process is considerably speeded up. However, a chip of rock not more than 8–10 mm in thickness can usually be broken from a hand specimen with a small hammer and then a thin section can be made.

The operations required to prepare a thin section after obtaining the fragment of rock are set out below.

Using 100 micron particle size (120 grade) carborundum abrasive, one surface of the rock fragment is ground flat on a piece of glass measuring about 30 cm × 30 cm and up to 1 cm in thickness; ordinary window glass is satisfactory if thicker glass is not available. Only a small amount of carborundum (half a teaspoonful), just moistened with water, is used for grinding. If too much water is present the carborundum tends to extrude from underneath the rock, and in consequence is much less effective for grinding.

After grinding with a rotary movement for about half a minute, the noise of the grinding changes because the carborundum grains lose their sharp cutting edges. The glass plate is washed clean and a fresh slurry of carborundum made on the plate. The time spent on grinding a flat surface will of course depend on how irregular the surface of the rock chip was to begin with.

When the surface of the rock is flat, the sample should be thoroughly cleaned with a jet of water before grinding with a finer grade of carborundum. The second stage of grinding should be carried out with 60 micron size (220 grade) carborundum and two periods of grinding, for about a minute each, with a fresh quantity of carborundum is all that is required at this stage.

After washing, a final grinding of one surface is made for about a minute with 12 micron size carborundum (3F grade). Again, after cleaning, the rock sample may be polished using cerium oxide (0.8 micron size) but this is not essential.

The next stage is to glue the smooth surface of the rock to a microscope slide in one of two ways. It can be achieved by using a cold-setting epoxy resin, which usually consists of two fluids which must be thoroughly mixed. The maker's instructions for using these should be followed carefully because these materials should not be allowed to come in contact with the skin and the vapour should not be inhaled. The refractive indices of epoxy resins vary but most are somewhat higher than the value of 1.54. For any work involving comparison of the refractive index of minerals with the mounting material, the refractive index of the cold resin should be ascertained. The chief disadvantage of using an epoxy resin is that it is very difficult to remove, if, for example, it became necessary to transfer the rock chip to another glass slide.

The alternative method is to use a material known as Lakeside 70C cement,[1] which is supplied in short rods and must be melted on a hotplate. This material begins to soften about 85 °C, so a hotplate which reaches 100 °C is quite suitable. A flat piece of aluminium or steel placed on a gas stove or on the element of an electric cooker at very low heat can be used for this stage, if no electric hotplate is available. A glass microscope slide and the rock specimen should both be heated on the hotplate until they are just too hot to touch, then some Lakeside cement is melted on the flat surfaces of the rock and the slide by touching them with the rod of Lakeside cement.

Whether the cold-setting epoxy resin or the Lakeside cement is used, the procedure is the same at this stage in that the flat surface of the rock chip must be attached to the glass slide with no air bubbles between the two surfaces. The rock chip is placed on the glass slide and, with a slight pressure and circular movement, the excess mounting material and air bubbles are squeezed out. The slide is then turned over to observe whether any air bubbles have been trapped between the rock and the slide; any bubbles must be gently extruded by pressure and, in the case of the Lakeside cement, this has to be done before the cement cools and becomes too viscous for the bubbles to escape easily. However, it can be reheated to render it fluid again. With the epoxy resin, since the hardening takes place over a period which depends on the variety, more time is available for extruding the air bubbles, but in this case the sample should not be heated because this only speeds up the hardening process.

If a diamond saw is available the rock fragment can now be cut from its original thickness of 5–10 mm to about 1 mm, otherwise it must be ground by hand. Its thickness should be reduced to about 0.2 mm (200 microns) using 100 micron size carborundum; at this thickness it is possible to see through the transparent minerals. Carborundum of 60 micron size should be used to reduce the thickness from 0.2 mm to 0.1 mm and at this stage quartz and feldspars should show bright second-order interference colours when examined under crossed polars.

The final stage of grinding from 0.1 mm to 0.03 mm is accomplished using 12 micron size carborundum. This is the stage in the whole process of section making which requires the most skill. The grinding has to be done very carefully to ensure that the section is of uniform thickness over its whole area, otherwise the edges tend to be ground preferentially and become too thin. The slide must be examined between each stage of grinding to check on the uniform reduction of the interference colours.

[1] *Lakeside cement is the proprietary name for a material manufactured in the USA and marked in the United Kingdom by Production Techniques Ltd., 11 Tavistock Road, Fleet, Hampshire.*

In the making of thin sections, it is generally assumed that the rock will contain some quartz or feldspar. These show first-order grey and white interference colours in a thin section of standard thickness and neither should show a first-order yellow or red colour. Thus a thin section in which quartz or feldspar shows colours in Newton's scale higher than first-order white is too thick. In making thin sections of limestones or evaporites where quartz and feldspar are absent, it is very difficult to estimate thickness; only an experienced thin section maker can do so accurately. With limestones, where the minerals show high-order interference colours, the section should be ground until sparite crystals and the internal structures of shell fragments are clear. Micrite will remain difficult to resolve even at high power.

It is usual to cover the section, either by painting the surface with a transparent cellulose lacquer, or better still with a glass cover slip as lacquer tends to scratch easily. This is done traditionally using Canada balsam diluted in xylene, but the process of heating the mixture at the correct temperature for the correct time requires some experience. We have found that it is quite satisfactory to fix the cover glass by either using the same epoxy resin which was used to attach the rock to the microscope slide, or by using a clear lacquer painted or sprayed onto the surface of the rock. As in the process of fixing the rock to the microscope slide, care must be taken to ensure that no air or gas bubbles are trapped between the cover glass and the rock. This is particularly important if the material has been applied by a spray, because some of the propellant may be dissolved in the clear lacquer. Any bubbles which are visible in the liquid after spraying should be allowed to burst before applying the cover slip. Only sufficient lacquer or Canada balsam to cover the slide with a thin layer of liquid should be applied.

The cover slip should touch the liquid on the slide at one end and be allowed to fall slowly onto the liquid. If any air bubbles are visible they can be extruded by gentle pressure on the cover glass. The excess lacquer or epoxy resin must be extruded to render it as thin as possible, otherwise the minerals cannot be brought into focus with a high-power lens because of the short working distance of lenses of magnification more than $\times 40$.

Finally when the mounting material has set hard, the excess can be scraped from round the edges of the cover glass with a razor blade or sharp knife.

Appendix 2

Staining a thin section of a limestone

The procedure detailed below, adapted from Dickson (1965), has been found generally satisfactory and has been used in preparation of most of the stained sections shown in this book. Two stains are required – Alizarin Red S and potassium ferricyanide.

1. Prepare a thin section of the rocks as described in Appendix 1 but omitting the coverslip. Ensure that no dirt or grease adheres to the surface.
2. Prepare two staining solutions:
   Solution A: Alizarin Red S – concentration of 0.2 g/100 ml of 1.5% hydrochloric acid (15 ml pure acid made up to 1 litre with water).
   Solution B: Potassium ferricyanide – concentration 2 g/100 ml of 1.5% hydrochloric acid.
3. Mix solutions A and B in the proportion 3 parts by volume of A to 2 parts of B.
4. Immerse the thin section in the mixture of solutions for 30–45 seconds, agitating gently for at least part of the time to remove gas bubbles from the surface.
5. Wash the stained section in running water for a few seconds.
6. Allow to dry.
7. Cover with polyurethane varnish or a coverslip in the normal way.

Note: The solution of Alizarin Red S in acid may be made up beforehand and will keep, but the potassium ferricyanide must be made fresh each time. A large number of sections can be stained with 250 ml of stain solution.

Appendix 3

Preparation of a stained acetate peel of a limestone

The following procedure has been found to work well with most lithified limestones of low porosity and has been used to make most of the peels shown in this book. Porous limestones should first be impregnated with resin, otherwise evaporation of the acetone will draw up water onto the stained surface after step 6 (below).

1. Prepare a slab of rock, grinding flat the surface to be peeled. The final grinding should be made using 3F grade carborundum powder.
2. Prepare stain solutions A and B in the concentrations described in Appendix 2.
3. Mix solutions in the proportion A:B, 3:2, and pour into a shallow container large enough to allow the whole of the ground surface to be in contact with the solution.
4. After ensuring that the surface to be peeled is free from dirt or grease, immerse the rock slab in the stain solution so that the surface to be peeled is completely covered by solution. This is done best by holding the specimen with the ground surface downwards, either by hand or in a clamp and retort stand, otherwise the solution will be wasted on the unprepared surfaces. The specimen should be immersed in the solution for 90 seconds. Agitate the solution occasionally to remove gas bubbles from the undersurface of the slab.
5. Rinse the stained surface with water and leave for a few minutes for excess water to evaporate.
6. Flood the surface with acetone allowing it to run off, taking the excess stain with it.
7. Cut out a piece of thin acetate sheet (0.003 inch thickness is suitable) slightly larger than the sample.
8. Arrange the rock sample with its stained surface uppermost and horizontal, taking care not to touch the surface.
9. Flood the surface with acetone.
10. Lower the acetate sheet onto the surface gently, taking care to expel any air bubbles which may have formed. No pressure is required.
11. Leave the slab and peel for half an hour at least, to allow the peel to harden.
12. Gently peel the acetate sheet from the sample.
13. Trim and mount immediately between two pieces of glass to keep the peel flat. Normal glass slides for thin sections are suitable for small samples.

The peel is now ready for examination under the microscope.

Note: (a) To make another peel of the same sample it is necessary to re-grind the surface with only the finest grit before repeating steps 4–13 above.

(b) The number of samples which can be peeled successfully using 500 ml of solution will depend on their surface areas. Using samples averaging about 5 cm square, 10–15 samples can be accommodated, although it will be necessary to increase the time in the solution as the acid becomes weaker. After 10–15 samples, the solution must either be discarded or strengthened with a 2–3 ml of concentrated hydrochloric acid.

(c) *All chemicals should be handled with great care. It is recommended that protective gloves are worn throughout the making of peels. Take care not to inhale the acetone fumes.*

References

Bathurst, R. G. C., 1975, *Carbonate Sediments and their Diagenesis. Elsevier, Amsterdam,* 2nd Edition.

Carver, R. E., 1971, *Procedures in Sedimentary Petrology.* Wiley-Interscience, New York.

Choquette, P. W. and Pray, L. C., 1970, Geologic nomenclature and classification of porosity in sedimentary carbonates. *Bull. Am. Assoc. Petrol. Geol.*, **54**, 207–50.

Dickson, J. A. D., 1965, A modified staining technique for carbonates in thin section. *Nature*, **205**, 587.

Dunham, R. J., 1962, Classification of carbonate rocks according to depositional texture. In W. E. Ham (Ed.), *Classification of carbonate rocks. Am. Assoc. Petrol. Geol. Mem. 1*, 108–21.

Flügel, E., 1982, *Microfacies analysis of limestones.* Springer, Berlin.

Folk, R. L., 1951, Stages of textural maturity in sedimentary rocks. *J. Sedim. Petrol.*, **21**, 127–30.

Folk, R. L., 1959, Practical petrographic classification of limestones. *Bull. Am. Assoc. Petrol. Geol.*, **43**, 1–38.

Folk, R. L., 1962, Spectral subdivision of limestone types. In W. E. Ham (Ed.), *Classification of carbonate rocks. Am. Assoc. Petrol. Geol. Mem. 1*, 62–84.

Folk, R. L., 1974, *Petrology of Sedimentary Rocks.* Hemphills, Austin, Texas.

Horowitz, H. S. and Potter, P. E., 1971, *Introductory Petrography of Fossils.* Springer, Berlin.

Johnson, J. H., 1961, *Limestone-building algae and algal limestones.* Colorado School of Mines, Golden, Colorado.

MacKenzie, W. S., Donaldson, C. H. and Guilford, C., 1982, *Atlas of igneous rocks and their textures.* Longman, Harlow.

Majewske, O. P., 1969, *Recognition of invertebrate fossil fragments in rocks and thin-sections. Int. Sed. Pet. Ser.* **13**, *101 pp.* Brill, Leiden.

Pettijohn, F. J. 1975, *Sedimentary Rocks.* Harper and Row, New York, 3rd Edition.

Pettijohn, F. J., Potter, P. E. and Siever, R. 1973, *Sand and Sandstone.* Springer, Berlin.

Scholle, P. A., 1978, *A color illustrated guide to carbonate rock constituents, textures, cements and porosites, Am. Assoc. Petrol. Geol. Mem.* 27.

Sorby, H. C., 1851, On the microscopical structure of the Calcareous Grit of the Yorkshire coast. *Q. J. geol. Soc. London*, **7**, 1–6.

Sorby, H. C., 1879, On the structure and origin of limestones. *Q. J. geol. Soc. London*, 35, 56–95.

Wray, J. L. 1977, *Calcareous algae.* Elsevier, Amsterdam.

Index

Plate numbers in bold type.
Page numbers, in lighter type, are given where there are no photographs, or where the discussion of the feature is on a different page to the plate. References to plates are given where illustrated material is described in text. Other photographs showing the same feature are mentioned at the end of each section of text.